...AKE
...ME
...ITY
FROM WIND,
WATER &
SUNSHINE

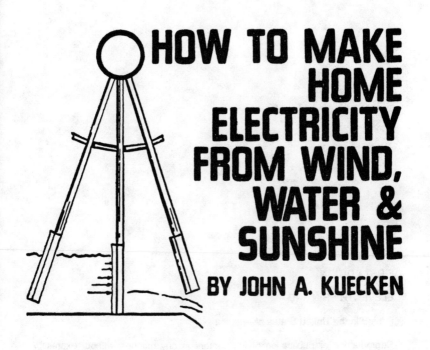

HOW TO MAKE HOME ELECTRICITY FROM WIND, WATER & SUNSHINE

BY JOHN A. KUECKEN

TAB TAB BOOKS Inc.

BLUE RIDGE SUMMIT, PA 17214

FIRST EDITION

ELEVENTH PRINTING

Printed in the United States of America

Library of Congress Cataloging in Publication Data

Kuecken, John A.
    How to make home electricity from wind, water, and
sunshine.

    Includes index.
    1. Electric power production.   I. Title.
TK1005.K83   621.312'1   79-12528
ISBN 0-8306-9785-3
ISBN 0-8306-1128-2 pbk.

Cover painting by Robert Barg.

Contents

Introduction

The invention of the water wheel and the windmill are lost in antiquity, but we know that these devices were widely used at the time of the middle ages. However, the wide use of the energy stored in combustible matter did not really begin until the development of the atmospheric steam engine by Newcomen in 1712 and the improvement of the steam engine by Watt in 1769. By 1815 Oliver Evans of Philadelphia had developed a high-pressure steam engine operating at 200 pounds per square inch (PSI). By the 1820's most major laboratories and some advanced manufacturing facilities were equipped with steam engines and line-shafts and belt drives. This was the beginning of the inexorable drive to replace the muscle power of men and animals with the tireless efforts of the machine.

The steam engine, the water wheel and the windmill were not to be the entire answer. Each of these machines was large, cumbersome and expensive, and the power that it developed could not be transported over any significant distance without great expense and losses. Yet another development was required. This came at first in the form of the demonstration of the dynamo-electric machine by Michael Faraday in 1831. Faraday demonstrated that electrical energy could be turned into mechanical energy and that an invisible electric current could also be made to rotate a shaft at high speed like a steam engine or a water wheel. A Frenchman by the name of Hypolite Pixii invented the machine in 1832 that was eventually to turn the course of history. The Pixii machine was a heteropolar permanent magnet commutated machine with an iron armature and a wire-wound core. It provided the first means to generate a continuous current from mechanical energy and freed the electrical experimenter from the requirement to generate electricity by burning expensive fuels like zinc, copper, iron and sulphuric acid in a chemical reaction.

The major significance of this was pointed out by Professor Siemens in 1851. At the speeds possible in those days, a one-inch diameter shaft could transmit perhaps one horsepower with a bearing every three feet. In less than a mile, the entire power would have been consumed in bearing friction, even with the finest bearings available. By comparison, the same one-inch diameter shaft made from copper could conduct over 2000 amperes of electricity. At 115 volts, this conductor could transmit 363 horsepower for a mile with very modest losses. In modern transmission line service at 220 kV, this same conductor would handle 695,000 horsepower. There is no torque; therefore, no bearings are required, and the conductor can be suspended from widespread poles or towers with only insulators to suspend it.

It was this ability to transmit energy easily and economically over great distances that was to eventually lead to the near elimination of physical drudgery by man and beast in most of the advanced nations of the world. In 1884 Edison opened the Pearl Street station in New York to sell electrical energy to operate the incandescent electric lamps he had invented and was then manufacturing. And the first central electrical power generating facility went "on line."

THE DIFFUSE ENERGY SYSTEM

For a period of nearly 85 years following the opening of the Pearl Street station, the consumption of energy and the size and power of the energy companies increased at a rate of nearly eight percent per year, doubling about every nine years with little real concern voiced by the public. Assuredly, there were periodic voices raised to note that the supply of fossil fuel was limited. As a matter of fact, a review of back issues of *Scientific American* reveals that about every other year some professor would calculate the year in which we would run out of coal or petroleum based on current rates of consumption, growth and known reserves. However, the public is not prone to worry about such matters and the expansion continued unabated.

The energy companies were less sanguine about the subject and began the push into nuclear energy as an answer to long term requirements for energy growth. This move has at times met with vigorous opposition from groups of people, some of whom oppose any nuclear installation with a ferocity that approaches religious zeal.

At the same time, the depredations wrought upon the state of our planet by the continuing consumption of energy in the form of fossil fuels or nuclear fuels has brought about a substantial public sentiment for the utilization of renewable resources for the satiation of our continually growing thirst for energy.

There are only two fundamental sources of energy available to mankind: the nuclear energy locked into certain elements in the death throes of long vanished stars and the radiant energy from the nuclear fires of our own sun. The latter is bequeathed on us in the form of the fossil remains of long-dead forests and in today's wind, rain, and sunshine. Arab oil embargoes and coal strikes have taught us that the fossil fuels are indeed a fallible source of energy. Indeed, the blackouts have taught us that the energy giants are also fallible. To certain people this suggests that the future may lie with the diffuse energy system in which energy generation would be widely scattered to harvest the energy sent us by the sun. The purpose of this book is to examine, at the practical level, those things which could be done to permit the individual to approach self-sufficiency in the generation of electrical energy.

This is a technical book and not a political book, and the attempt is made to provide practical technical suggestions regarding the generation of home electricity. It is down-to-earth to the extent of suggesting the conversion and use of junkyard parts to achieve this end—where such usage is practical.

The book attempts to treat the system and design considerations in a straightforward and practical manner with minimal resort to theoretical or highly mathematical discussions or theories. Wherever feasible, simple rule-of-thumb design techniques are presented. The subject matter of the book is at times sophisticated and requires some mathematics for understanding. In places where this occurs, every effort has been made to provide worked-out examples in the interest of maximum clarity. Understanding of the situation may also require some background in fundamental electricity. For the reader who does not have this background, we recommend resorting to one of the numerous texts on the subject.

Again, from the viewpoint of practicality, wherever possible the economics of any given approach are considered. This extends to the fact that the overwhelming majority of the consideration in the book is devoted to the production of 115-volt 60-Hz electricity despite the fact that low-voltage DC is much easier to produce and understand. Unfortunately, it is also much more expensive to use. The overwhelming majority of the devices, appliances and motors used in the home will operate only on 115-volt 60-Hz electricity.

While it is possible to purchase a refrigerator and perhaps an automatic washer which runs on low-voltage DC, these items are prohibitively expensive to purchase and difficult to have serviced. The text principally discusses the construction of installations capable of handling practical household electrical loads of existing and available appliances. Considerable attention is devoted to optimizing the match of the system to the household requirements to obtain maximum benefits of convenience with minimum cost and effort.

John A. Kuecken

Home Electrical Requirements

A logical place to start this book is with a consideration of just how much electricity one is likely to need in a home. If one is seriously considering the installation of an independent electrical plant, the proper sizing of the installation is of considerable importance. If one wanted only enough electricity to operate a transistor radio or perhaps an emergency lamp, the solution can be achieved very simply with a few solar cells and a storage battery. However, if you intend to operate an electric automatic drier and an air conditioner along with the usual household appliances, a considerably more complex and expensive installation is required.

A fair place to begin, assuming that your home is presently electrified, is to survey your electric bills. The power company prints the number of kilowatt hours on the bill. You will find that the examination of the year's bills will show a certain amount of seasonal fluctuation so it is best to consider the billing for the entire year. I live in a fairly typical suburban home in a cold northern climate and at 47° North latitude. The days get relatively short during December, so the lights are burned more often and for longer periods. In addition, the furnace blower runs more frequently. Added to the usual lights we have the following appliances:

Small refrigerator
Large self-defrosting refrigerator/freezer
22-cu. ft. freezer
Steam iron

Automatic dishwasher
Automatic clothes washer
Automatic electric dryer
Gas furnace with electric blower
Electric stove
Small electric oven
Disposal
Hair dryer
Large color TV (solid-state)
Small color TV (tube)
Stereo (solid state)
The usual assortment of radios, mixers, etc.

We do not have an air conditioner and do not heat water electrically. However, this is somewhat offset by the presence of a fairly extensive electronics lab and machine shop. The home is generally occupied by two people during the college year, and the population can rise to six or so during holidays.

In a typical year I find that we consume about 10,200 kilowatt-hours (thousands of watt hours). Because there are 8766 hours in the average year, this works out to an average electric load of 1.164 kilowatts. The peak month is February. At a rate of five cents per kilowatt-hour (kWh hereafter), we would pay an annual electric bill of $510.

ECONOMICS

According to some power company figures, the price of electrical energy has been growing at the rate of eight percent per year. Therefore, at the end of 10 years I will expect to have paid the power company a total of $7388 for electric service. It is obvious that I could afford a reasonably sized investment for power generating equipment which operates from a "free" energy source. However, to be fair we have to assume that if I currently have the money it would be earning something in an investment or savings account. Assume the interest rate to be six percent on a long-term account. With this correction, we find that our break-even investment would be $4,125.50, for this sum, if invested at six percent interest, would grow to $7388 at the end of 10 years.

Naturally, even though the prime energy may be "free," any continuing cost to maintain the equipment would have to be deducted from the original investment for a "break-even" situation. If this maintenance cost were as little as an average of $100 per year, we find that the break-even initial investment could be no higher than $4,125.50– 736.01 or $3,389.49. The reason we did not deduct $1000 is the fact that the sum of $736.01 invested at six percent would permit us to withdraw $100 at the end of every year for 10 years to provide the money for maintaining the equipment. As we shall see shortly, this is a fairly difficult requirement to meet.

Naturally, there are a variety of reasons why one would want to manufacture his own home electricity. He might live on an island or in an area which has not been electrified. He might also be concerned about the environment and consider that a "diffuse" energy structure is preferable to the present "centralized" energy structure. He might also have doubts about the future ability of the existing energy structure to furnish him with the required energy. The reason for this brief sally into the economics of the situation is to provide a comparison standard regarding the economic realities of the problem. Even though the prime energy source comes to us "free" in the form of wind, water, or sunshine, an engineer would be remiss if he did not discuss the costs of acquisition, maintenance and replacement of the necessary equipment. Nature provides us no free lunches! As we proceed through some of the individual techniques and devices to be described in subsequent chapters we shall attempt to keep an eye on the economics of the situation in order to have some price index which would be paid for each.

PEAK POWER LOADS

The yearly average power found on the power company bills tells only a part of the story and that part is principally economics. The monthly peak power is generally reached in February, which is normally the coldest month in our northern climate. During this period, power consumption rises to about 960 kWh/mo or an average load of 1.314 kW. During February, the solar input is minimal, and it might be difficult to keep a small hydro plant operating since small bodies of water will freeze to a depth of a meter or more. A wind-operated plant would have a better chance for success provided that it could be kept from icing up.

Naturally, these comments do not apply to locations like Tucson, Arizona or the Florida Keys. However, the point remains the same: The wind does not always blow, the water may not always flow, and the sun certainly does not always shine, at least not at night. In order for a home electric system to be completely self-sufficient, it must be provided with enough capacity to store energy to carry at least a reduced load during periods when the "free" prime energy resource temporarily disappears. In certain climes it may be advisable to provide alternate access to prime energy with both a wind and water or a wind and solar installation or some other combination. For example, a solar steam installation could be built to include a conventional boiler which would burn some renewable-resource fuel such as wood or cornstalks. This action does carry the burden of introducing combustion products into the atmosphere.

To put the matter into perspective, a very efficient gasoline engine consumes 0.6 pounds of fuel per horsepower hour. There are 745.7 watts in one horsepower, so our 1.3 kW represents 1.76 horsepower. A small wood-burning steam plant would be no more than one-fifth as efficient and could easily consume 5 to 10 times this weight in wood or cornstalks or other low-grade fuel per horsepower hour. This translates into a fuel consumption of between 5¼ and 10½ pounds of fuel per hour to maintain our 1.3 kW electrical load. One month without any significant sunshine would require the consumption of between 3800 and 7700 pounds of low-grade fuel. With typical hardwoods, this would represent a woodpile in a cube five feet to seven feet on a side.

A side benefit of a solar steam system would be the fact that the "waste" heat from the condenser would be capable of providing a good share of the heating requirements of a rather large-sized house. This would be true regardless of whether the system were operating on sunshine or the auxiliary fuel. The "waste" heat capability is an advantage not accruing to either a falling water or a wind-driven plant and would tend to offset the naturally higher cost of a solar steam installation. In our discussion of solar steam installations, we will examine this facet in more detail.

In a hot, sunny clime such as the American Southwest, the outlook for a solar steam installation for electrical energy is much more attractive. First of all, the sun shines more often and secondly, the peak system load will generally occur due to air conditioning that is mostly needed when the sun is shining. In many of the desert locations the night air is rather cool, and little if any air conditioning is required at night. In such locations, the principal night time loads can probably be reduced to food refrigeration and lighting.

In any event the maximum average load coupled with an estimate of the duration for which this load must be maintained represents a measure of the energy storage capability which must be incorporated into the system. Probably the only system which could reliably operate without some energy storage is a hydro installation in an area where the water supply could be depended upon not to dry up or to freeze.

PEAK CURRENT LOADS

The watt-hour meter installed in your home is deliberately designed to register only the actual watt-hours of electricity consumed. This is the actual heating value of the electricity. It is the equivalent of the amount of direct current which would have been required to produce the same amount of heat in a resistor. However, the actual electricity used in nearly all homes is alternating current and for alternating current power is determined by voltage, current *and* phase angle:

$$\text{Watts} = \text{Volts} \times \text{Amperes} \times \cos \theta$$

where: θ is the phase angle between the voltage and the current. Stated another way, the product of the voltage and current entering your electrical system does *not* equal power until corrected by the phase angle. Since the numerical value of cos θ varies between one at 0° and 180° and zero at 90° and 270° the current will nearly always be greater than watts divided by volts. Furthermore, since the watt-hour meter provides only average readings on the dial, the dial reading can tell us nothing about the actual current draw at any instant in time. The speed of revolution of the watt-hour meter disc will tell us about the rate of *power* consumption, but it corrects out the phase angle. Therefore, it will *not* tell us about the current draw. A lightly loaded induction motor can have a phase angle of up to 75°. Because the cosine of 75°=0.269, the actual current being drawn from the line is 3.864 times as great as watts divided by voltage. The generator naturally must supply both the voltage and the current even though there is little actual power being consumed. It is this very short term voltage/current rating that determines the size of any alternating current machine and the size of the prime mover required to drive it. For this reason you will find that a typical alternating current generator will be rated in thousands of volt-amperes (kVA) rather than Kilowatts (kW). The plant kVA rating will be a number that you will have to learn in order to construct or specify the right size for your installation.

A rough top limit for your electrical system can be obtained simply by looking at the main fuses in your fuse box. My fuse box contains a pair of 60-ampere fuses through which all of the electricity used in the house must pass. The system is a 220-volt split-phase system with 110 volts each side of ground. Figure 1-1 shows this arrangement. Since neither of these fuses has ever blown in a number of years of service, it is safe to assume that the system has never drawn currents in excess of the capabilities of these fuses. The

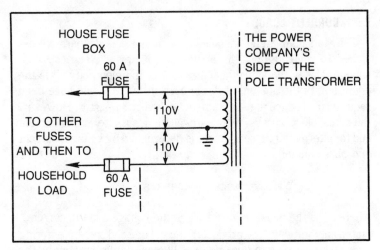

Fig. 1-1. Electrical service installation.

fuses are rated to carry 60 amperes continuously and 120 amperes for one minute. Thus, it is safe to assume that we have never continuously exceeded 220 volts times 60 amperes = 13.2 kVA nor exceeded 26.4 kVA for a period of one minute. Obviously this is a very large rating compared to the 1.3 kW averaged in the peak month of February.

The reason for this large safety factor lies mainly in the starting characteristics of induction motors. A number of the appliances around the house, such as the furnace blower, the electric clothes dryer, the automatic clothes washer, will draw currents in excess of 25 amperes when they start. Once up to speed, the current will fall back to something more like five to seven amperes. However, nothing in the existing system prevents five or six such machines from starting simultaneously. In this case the rating of the main fuses is approached or even momentarily exceeded. The power company also has to cope with this problem of transient starting loads. However, they have a very large number of customers and the probability that all of the motors they serve starting at the same instant is negligible except in the fire-up operation after an interruption of service. In order to avoid having something like a 5:1 or 10:1 ratio between installed kVA rating and average power sale they do two things:

- •After any system shutdown, the system is brought up one section at a time and the next section is not switched in until the first has been stabilized.
- •The power company protects itself with *power factor correction* which minimizes the value of the phase angle seen at their

installation. This power factor correction provides for the minimum number of delivered amperes required to provide the actual load wattage.

In order to provide anything like a reasonable and economic ratio between average electrical usage and installed kVA rating for the plant, a few measures must be taken to ensure some of the same capabilities in a home electrical system. Obviously certain measures could be used to prevent all of the motors from starting simultaneously. With available electronics, it is much less expensive to provide a sophisticated control than it is to construct a system with 5 or 10 times the installed kVA rating. We will also eventually discuss systems which are tolerant to short duration overloads.

The overload problem is especially pernicious in a small system since some systems may tend to be severely slow. If the overload occurs because two motors are trying to start simultaneously, for example the refrigerator and the furnace blower, the overload is likely to reduce the system voltage to a level where neither machine can come off of the starting winding and start. In this case the motors will draw a very large current and will usually burn out within a minute or so if not protected by a thermal cutout or fuse. The fusebox fuse will seldom protect a stalled motor from burnout since the starting winding will often burn out if it carries a current not much more than the normal operating current in the run winding. To avoid this type of failure, a great many appliances have a built-in thermal cutout which opens when the windings start to get hot. This cutout is usually not built to withstand frequent usage since it is only an emergency facility. It is far better to provide your system with some other form of protection against overload and undervoltage.

MEASURING SYSTEM REQUIREMENTS

If one were starting entirely from scratch to build a home and to buy every one of the appliances new, it might be possible to obtain complete data on the starting and running currents for each of the appliances and do the entire system design on paper. One would have to contact the appliance manufacturer's engineering department and perhaps forgo purchasing appliances for which the data was not furnished. However, it is highly unlikely that the salesman in a discount appliance store will be able to furnish you with this information, or for that matter even be able to understand the question. It therefore seems that the home electrifier will have to perform some measurements for himself in most cases.

The illustration in Fig. 1-2 shows three common methods of measuring the AC line current. A standard AC ammeter which

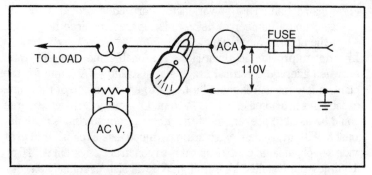

Fig. 1-2. Three types of ammeters.

connects in series with the lines can be purchased from any number of electrical suppliers. Somewhat easier to use in some cases is the clamp-on ammeter which is simply clamped over insulated wires. Also shown is the current-transformer/AC voltmeter technique. Note that the measurement is made on the "hot" side of the line. Always pull the fuse before connecting an ammeter in series with the line. Never connect an ammeter across the line or between points that show any significant voltage drop.

Failure to observe the first caution will get you shocked, burned or electrocuted. Failure to observe the second will instantaneously destroy the instrument in a ball of fire.

It should be noted that most modern homes are wired with composite cable carrying the "hot" wire (usually black), a ground return wire (usually white), and a safety ground wire which may be either green or bare. In order to obtain a reading with the clamp-on ammeter it is necessary to clamp it over *only* the black wire. If all three wires pass through the clamp window, the reading will be zero regardless of the current. The clamp-on ammeter may be safely clamped on an energized wire *only* if the insulation is intact and there are no exposed terminals or bus bars in the vicinity.

Both the standard ammeter and the current transformer require that the circuit be de-energized by removing the main fuses. The "hot" line is then broken and the instrument connected in series. The circuit is then re-energized, and the current reading taken.

The current transformer may be purchased from an electrical hardware supplier. However, a very acceptable one may be constructed by re-winding a small step-down transformer. In particular, I have used a transformer which had a 115V primary winding and a 24-volt 1.3-ampere secondary. Units of this rating may be purchased at a local radio store or perhaps found in the junkbox. The

18

case/clamp on this unit was pried open and the laminations removed one at a time and carefully saved.

The laminations are held together only with varnish, and the "I" pieces can easily be split off with a putty knife. The associated "E" section can then be split loose and removed. When the core is completely off, the secondary winding (which is usually on the outside) can be stripped off. If the primary winding is intact, it may be left on the bobbin. For a transformer of about this rating, namely 31 volt-amperes, the windings are usually run at 8 to 9 turns per volt. An eight-turn secondary can then be wound on the bobbin using #10 B&S gauge insulated magnet wire. Be careful to ensure that the secondary is well insulated from the primary and from the core eventually. This can be done with certain types of tape. However, conventional masking tape in several layers is adequate if the transformer is not allowed to run too hot. "Too hot" in this case would be about 130°F or where the unit would be uncomfortable to touch.

At this point the reassembly of the core can begin. It is important to resist the temptation to make all the "E" laminations face the same direction. It does make reassembly easier, but it also makes for a very poor transformer. Also try to make sure that all of the laminations get back in. The easiest technique is to put in three or four "E's" facing in alternate directions and then slip the necessary "I's" between them. When these are all in place, the core cover may be clamped around them and the retaining tabs hammered back down while the unit is clamped in a vise. A little epoxy inside the core cover will not hurt and may help prevent the unit from buzzing.

If these steps have been followed carefully, you should now have a transformer with a 110-volt primary and a 1-volt 30-ampere secondary. The secondary winding will actually withstand considerably higher currents for short intervals. The next step is to connect the original primary to 110 volts and measure the primary and secondary voltages with no load on the secondary. The primary probably consists of about 880 turns of No. 30 wire. The ratio of the two voltages may be used to find the current transformation ratio:

$$\frac{\text{Voltage Primary}}{\text{Voltage Secondary}} = \frac{\text{CurrentSecondary}}{\text{Current Primary}} = \frac{\text{Turns Primary}}{\text{Turns Secondary.}}$$

Now we come to the purpose of the resistor "R" shown in Fig. 1-2. Let us suppose that we would like to arrange the unit so that 30 amperes flowing through the heavy winding would give us a 30-volt output on the light winding. The current in the light winding should be 30A/110 or 0.273 amperes. If we force this to flow through a

110-ohm resistor there will be a 30-volt drop across the resistor. The resistor will be dissipating a power of:

$$\text{Power} = i^2 R \text{ or } (0.273)^2 \times 110$$
$$= 8.2 \text{ watts}$$

The 273 milliamperes in the light winding is still within reasonable limits, and the unit should run relatively cool. An ordinary multimeter with AC scales would be adequate to measure the voltage on a 1V/ampere basis.

One of the advantages of the current transformer is the fact that it can be used to determine the waveshape and the phase angle of the current being drawn. The waveshape is important since a great many modern devices employ silicon control rectifiers (SCRs) in a phase-control circuit to regulate the current being drawn by a controlled appliance. These units tend to mutilate the waveshape. The mutilated waveshape will not read properly in an ordinary ammeter.

POWER FACTOR CORRECTION

We had earlier mentioned the fact that the power companies tend to protect themselves by correcting the power factor of their system. This is done for a variety of reasons, including voltage regulation. However, it is usually done to keep the cos θ term close to unity so that they have a kVA/kW ratio close to unity. The system losses rise as the square of the system current and a power factor of unity assures that current in the system is minimal for the actual number of watts being consumed.

To try to bring this point home I will refer to a real physical example, the motor which drives the drill press in my shop. This is an elderly ball-bearing unit that drives the drill press quill which runs very freely. The drill press is also equipped with a 100-watt lamp. Figure 1-3 shows the oscillograms of the units with a scale change. It can be seen that both units distort the waveform from a pure sinewave, and that the lamp current leads the voltage slightly and the motor current lags the voltage by quite a bit. Now the motor is absorbing 7.2 amperes peak $\times$ 0.707 = 5.09 amperes rms and the line voltage is 163V peak $\times$ 0.707 = 115V rms for a total of 115 $\times$ 5.09 = 586.62 volt amperes. However, the actual power being drawn is:

$$115V \times 5.09 A \times \cos 84.5° = 56.23 \text{ watts.}$$

The disparity between the volt amperes and the watts is a ratio greater than ten.

Since the power dissipated in the generator winding is proportional to i^2, we see that our generator and wiring is dissipating over a 100 times more than is necessary in terms of the work being accomplished.

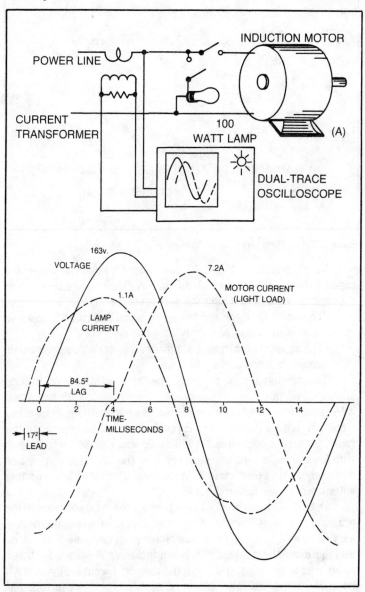

Fig. 1-3. If the oscilloscope were connected as in A, the oscillograms in B would appear.

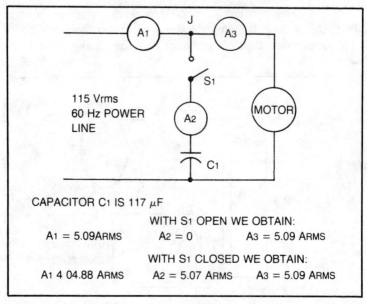

Fig. 1-4. Determining power factor correction.

Now referring to Fig. 1-4, we see that a large condenser (or capacitor) has been added with a switch and three ammeters. From the switch open/closed data we see several things:

- The addition of the condenser cuts the line current down to exactly the value to supply only the real work.
- The sum of the currents at junction J is not zero as we would expect it to be in the DC case.

The advantages of reducing the current in the generating apparatus by this magnitude is relatively obvious. With any given kVA rating of generator we could run more than 10 times as many induction motors if they were carefully tuned out to eliminate the reactive current component. In the following chapter, which deals with load management, we shall see how this can be accomplished with either an approximate solution or a continuously correcting automated power-factor device.

Our intention in this book is to present practical solutions to the practical problems of generating home electricity. It is not necessary to be able to calculate power factor correction elements to do this therefore the mathematics is not included. However for those interested in delving further into the matter, I would suggest TAB book No. 929, *Solid-State Motor Controls*. This text in turn has references to a number of other more mathematically oriented texts.

Motors and Generators

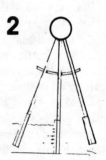

2

The basic principle upon which nearly all electric motors and generators operate is the discovery by Ampere that a wire carrying an electric current experiences a physical force when it is immersed in a magnetic field and the discovery by Oersted that a wire moved through a magnetic field develops a voltage. These actions are complementary; that is if the moving wire is connected to an external circuit so that the induced voltage can force a current flow, it will be found that the wire develops a mechanical force which tends to resist the motion. In other words, it becomes necessary to do mechanical work to move the wire.

The converse principle also works. If a current is forced into a wire in a magnetic field, the wire will develop a force which tends to cause it to move. If nothing opposed this motion, the wire would accelerate to a speed where the induced voltage would exactly buck out the external voltage source and would bring the current to zero. If on the other hand, something opposes the motion of the wire like an external mechanical load or the inertia of the wire, then the wire would accelerate to a speed where it bucked out a portion of the external voltage and only enough current would flow to accomplish the mechanical work being done. This bucking voltage is termed a *counterelectromotive force* (counter emf) and is responsible for the workable nature of the electric motor.

While we are not interested in learning the details of designing electric motors in this text, we should examine in a qualitative way some of these principles so that we may understand some of the phenomena which are encountered in a small electrical system. Figure 2-1 shows the simplest kind of electric motor or generator. A coil of wire having a number of turns is pivoted so that it can rotate between the poles of a magnet. Beneath this is a series of further simplified figures showing a sort of moving picture reduced to only the wires themselves.

First of all, if the coil is rotating at some constant speed ω we can see that the coil sides A and B will cut the magnetic lines β at the fastest rate when the coils are in the positions designated 1 and 5 and will not be cutting any lines at all when in position 3. Furthermore, wire A will be cutting the lines in the opposite direction at position 5 as it was at position 1. Therefore, we would expect that the voltages induced in A and B would be reversed. If the field between the magnet poles is uniform, it can be shown that the induced voltage is proportional to the sine of the angle α since this is the relationship which determines the rate at which the magnetic lines are cut. The machine is obviously producing alternating current of a frequency ω equal to the mechanical rotation speed of the shaft.

If the multiple turns of the coil are wound in the form of a simple helix it is fairly easy to visualize the fact that the voltages will add together. The voltage induced will be in direct proportion to:
• The strength of the magnetic field
• The rate of rotation, ω, and the radius of the rotor
• The number of turns on the coil
• The sin of angle α (which varies with time).
If we wanted to use this machine for either a motor or a generator, it would be necessary to provide some means of connecting terminals A and B to the outside world. If we wanted alternating current operation, we could make the rotary connection to the outside world via slip-rings as shown in Fig. 2-2.

The commutator bars of Fig. 2-3 not only provide the sliding rotary connection but they also permit the outside connection to "swap hands" between terminals A and B so that brush N could always be positive and brush M could always be negative. That is, if the insulating bar is arranged to be vertical when the coil is horizontal, and the brushes were arranged roughly horizontal. We shall explain the term "roughly" in short order.

There are several things of importance to be noted from the above conclusions. First and foremost is the fact that with any given machine, the faster it turns, the higher the output voltage will be. Now the current carrying capacity of the machine is largely determined by the size of the wire with which the machine is wound. Therefore, the faster the machine turns, the higher the available power. In aircraft usage, where weight is at a premium, advantage is taken of this fact. A bipolar alternator for aircraft use is generally rotated at 400 rev/sec or 24,000 rpm in order to get the maximum output from the smallest and lightest machine. A given size aircraft alternator with 400-Hz output will generally produce 400/60 times as much power as a 60-Hz machine of the same size. Unfortunately the machines are a good deal more expensive since they must be capable of very high speeds without bursting from centrifugal force, and the armature iron must be made of 0.004-inch laminations rather than the 0.014 inch used in the 60 Hz machine. This high speed advantage holds for motors and transformers as well. A 400-Hz transformer requires only about one-seventh as much iron as a

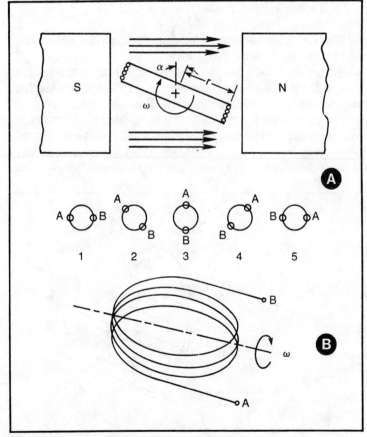

Fig. 2-1. In A, the simplified motor or generator. In B, if the wires of the coil are wound in a simple helix, it is fairly easy to see that the voltages in the two sides and the successive turns add up.

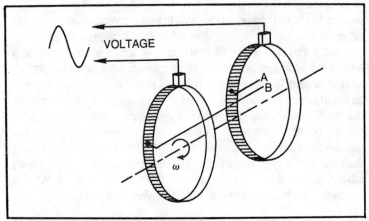

Fig. 2-2. The slip ring for AC connection.

60-Hz design with the same VA rating but again the laminations must be thin to avoid excessive losses.

Four-pole 12,000 rpm or eight-pole 6,000 rpm machines can often be found at very low prices on the surplus market, and the machines are outstanding performers. However, they cannot be used to run most 60-Hz appliances that have either transformers or motors. An exception is the type of motor used in most single-speed hand tools such as saws, hand drills, etc. These will run successfully on the high-frequency current, as will incandescent lamps. With

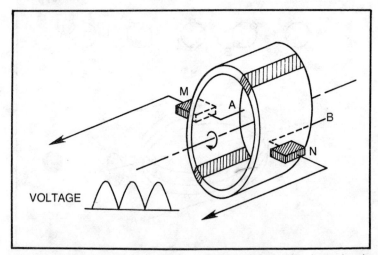

Fig. 2-3. The commutator (simplified) permits the outside world to change hands with terminals A and B so that a unipolar voltage is presented on the machine terminals.

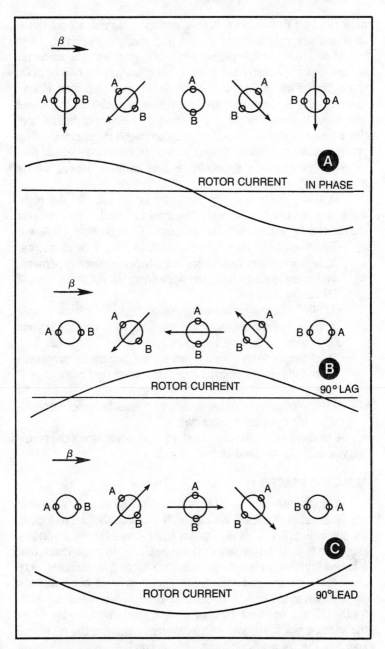

Fig. 2-4. Armature reaction with the rotor winding simplified to just two windings. At A, the machine is connected to an external resistive load. At B, the external load is lagging 90° as if it were a very lightly loaded motor or inductor. At C, the external load is a capacitor so the output current leads the output voltage by 90°.

most radios, TVs, washers, dryers, and such, the unit will not run at all or will severely overheat on the high-frequency current.

If we are to directly generate 60-Hz current with the machine, we are generally restricted to 60 revolutions per second or 3600 rpm in bipolar machines and 1800 rpm in four-pole machines. Unfortunately, this is a notable restriction. You can see that the speed with which the machine cuts the flux is also a function of the radius of the machine so that one would expect that the 60-Hz machine will be pretty sizeable and indeed this is the case. In most cases I doubt that even a very strong man would be able to lift a 5-kVA, 60-Hz alternator off the floor!

At this point you are probably about to ask, "If the high-frequency devices have so much advantage, why don't we convert the electrical system to 400-Hz operation? The answer is as follows:

- Because of the high speeds and high frequencies; motors, transformers and gear trains are inherently more expensive. Wider usage would bring the price down but not to the level of 60 Hz components.
- At 60 Hz the transmission line effects begin to be troublesome at 100 miles and become very complicated to manage beyond 350 miles. At 400 Hz the limits reduce to 17 miles and 52.5 miles respectively. An upward shift in the frequency would make the power company interties a great deal more expensive.
- Design of power-station size components becomes very difficult for the high frequencies.

As we shall subsequently see, modern electronics offers the possibility of utilizing the best of both worlds.

ARMATURE REACTION

It was noted earlier that in our machine of Fig. 2-1, we would set the brushes "roughly" horizontal. To explain this, let us look at the top of Fig. 2-4. This illustration again shows the rotor winding simplified to just two terminals. It can be shown by experiment that a helical winding with a current flowing through it behaves very much like a bar magnet with the magnetic strength being directly proportional to the amount of current. We have also seen that even for the DC machine with the commutator, the voltage developed by the winding is a sine wave with a frequency equal to the rotational speed of the rotor in revolutions per second.

The top illustration shows what happens if the machine were connected to an external resistive load. With this condition, we can see that the current flowing through the rotor tends to generate a magnetic field which fluctuates twice per revolution and is at right angles to the machine flux. The combination of this flux and the machine flux tends to shift the net direction of the flux in the machine so that in the DC machine case the machine would produce more output if the brushes were offset to align with the next flux field. In modern DC machines this effect is offset by other techniques. However, let us see what this effect does in an alternator.

The next illustration down shows the effect when the external load is lagging 90° as if the external load were a very lightly loaded motor or inductor. It may be seen that the net effect of the armature current is in a direction to directly oppose the magnetic field of the machine. Any reduction in this field will tend to directly reduce the generated voltage. If the field magnetism is not increased the output voltage will sag seriously.

The next figure below shows the effect of the armature reaction when the external load is a capacitor so that the output current leads the output voltage by 90°. Here it may be seen that the action is to strengthen the machine field. This in turn would tend to make the output voltage climb and in some cases soar. This is one of the problems associated with energizing a long unloaded transmission line. The line will act as a large capacitor, and the output voltage of the machine may climb beyond the ability of the field excitation system to control it. Note that this effect is a positive feedback or runaway situation. The higher the output current is, the greater the output voltage is, which in turn increases the output current. The effect is limited by the saturation of the system iron, after which the magnetism increases slowly.

For the inductive load, the armature reaction effect is less disastrous since an increase in current decreases the voltage which further decreases the current. Unfortunately, the undervoltage can also burn out motors since they may not be able to start and thus may draw excessive current.

This armature reaction is one of the most serious reasons why care should be taken to keep the power factor up in a small system. Unlike the power company, the home electrical system does not have a very large number of independent loads to average out. In some instances, a single motor going into the start cycle may prove to be a major part of the system load.

A MORE PRACTICAL MACHINE

Figure 2-5 shows a somewhat more practical form of AC motor or generator. If winding AB were energized with a direct current, the rotor or armature becomes a two-pole magnet. As it is rotated, it induces a voltage in windings CD and EF. Note that the sine wave voltages introduced in these windings is in quadrature, or the voltages are 90° out of phase: voltage CD is zero when voltage EF is maximum and vice versa. The rotating magnetic field principle was invented by Nicola Tesla and forms the basis for all induction motors.

With DC excitation on the rotor and if the machine were in synchronism with the line and the voltages CD and EF applied from the outside, this machine would run as a synchronous motor and could deliver power through the shaft. Unfortunately, in this form the motor has no starting torque and would have to be brought up to synchronous speed by some other means.

If winding AB were shorted at terminals A and B and the voltages CD and EF applied from an external source, the device becomes a two-phase induction motor. The quadrature excitation of the two phases produces a rotating magnetic field. This field in turn induces currents in the shorted windings of the rotor which tend to drag the rotor along with the rotating magnetic field. This type of motor will develop a very sizeable starting torque and will run at some speed lower than the line.

If the rotor were turning at precisely the line speed, say 1800 rpm, there would be no voltage induced in the rotor winding and the motor would develop no torque. As the rotor slows down, more and more current is induced in the rotor winding by the slip frequency between the line and the rotor, and the motor develops progressively higher torque. This ability to slow a little and then develop more torque makes the induction motor a very useful type and accounts for its high popularity in home appliances. The curve of Fig. 2-6 illustrates the speed/torque curve for an induction motor.

The induction motor of Fig. 2-5 would continue to run very well if it had first come up to speed and the voltage was removed from one of the windings, CD or EF. Once it is near operating speed the flywheel effect will keep it synchronized with the now pulsating but *not* rotating magnetic field. At the start, if only a single winding is energized this motor develops no starting torque. All induction motors start as polyphase devices. Once started, they may continue to run as single-phase devices. For some motors, the phase shift for the other winding is obtained by placing a capacitor in series with the winding. For the more common type, the phase shift is obtained from the design of the winding. In this case, however, the design of the starting winding is such that it cannot be continuously left across the line without burning up. To protect the starting winding, a centrifugal switch is provided to interrupt the current to the starting

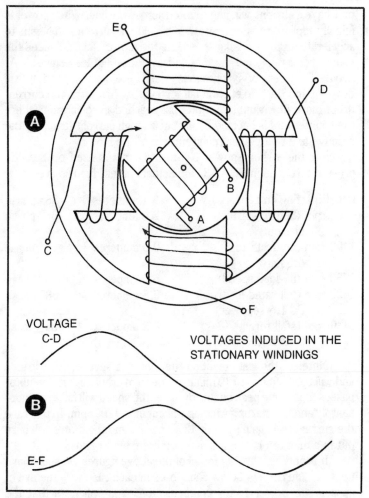

Fig. 2-5. A more practical AC motor at A, and the voltages induced in the stationary windings at B.

winding when the motor exceeds about 65 percent of the synchronous speed. It is this winding that usually burns in an undervoltage condition. If the motor does not get up to the trip speed, then this switch does not open to protect the starting winding.

STARTING CHARACTERISTICS

Nearly all motors have the same basic pattern of starting characteristics. With the motor at rest, it is developing no counter emf to buck out the supply voltage. In addition, the induction motor may have a starting winding shunted across the line. When power is initially applied, the motor draws a big slug of current and begins to accelerate up to operating speed. As the motor picks up speed the counter emf rises and the motor draws less and less current. At about 65 percent of synchronous speed, the starting winding will drop out and the current will fall further. At this point the current drops more slowly until the motor has reached an equilibrium where the counter emf is bucking out all but enough current to supply the reactive and real power requirements of the motor.

For the ⅓-horsepower ball-bearing motor used on the drill press referenced in Fig. 1-3, the actual current figures are:

Initial Starting current	19.1 amperes	78° phase
1400 rpm (before starter dropout)	12.0 amperes	65° phase
1400 rpm (after starter dropout)	9.0 amperes	58° phase
1750 rpm (no-load speed)	5.0 amperes	84.5° phase
1620 rpm (full rated load) (⅓ Horsepower)	5.5 amperes	68° phase
1440 rpm (stall torque 0.73 Horsepower)	6.0 amperes	38° phase

Note: In the last condition the motor is severely overloaded and will burn out in 5 to 10 minutes due to overheating. If the motor passes the torque peak on the low side, the speed will fall even more sharply and the starting winding will cut in at 1200 rpm. In this case the current will again jump to 12 amperes, and the motor will burn out in a minute or so if it is already hot from overload.

It is noteworthy that for all of the above figures we could have kept the line current below 5 amperes in each case with the use of appropriate power factor compensation. You will note that the compensation would have to be dynamically changing since this action is quite fast. With the light load of the drill press, which spins very freely when not actually drilling, the motor will come up to speed in about three seconds and it drops off the starting winding at about two. However, the motor on our automatic washer is another matter. It can take up to 15 or 20 seconds to drop off the starting winding when it is accelerating a full rated load of clothes and water on the spin cycle. This begins to tax the rating of a 15-ampere fuse if there is anything else on the line simultaneously.

For the small self-sustaining electrical system, the thing to look for in establishing the required short-term rating of the system is the current draw of the appliances which have a large inertial load to run:

the automatic washer which must accelerate a heavy tub full of water and clothes, the furnace blower which has to accelerate a large and heavy squirrel-cage fan, the refrigerator, etc. these are the items which tend to draw a very heavy current during the starting cycle.

The capacitor start motor and the capacitor start/capacitor run motor usually represent a somewhat better power factor than the usual starting-winding- induction motor. These units also tend to be more expensive. In certain applications such as the furnace blower it is sometimes possible to purchase and substitute one of these units. However, in the majority of household applications, such as refrigerator, automatic clothes washer, and automatic dishwasher, the motor is installed as an integral part of the unit and the substitution of another unit with a better power factor is not practical.

THE INCANDESCENT LAMP

Before leaving the subject of starting surges it is worthwhile to note that most resistive type heating elements change resistance considerably in the course of heating up. The 100-watt lamp on the drill press for example will draw an initial current of 10.8 amperes at the first instant that it is connected across the line. This rapidly falls

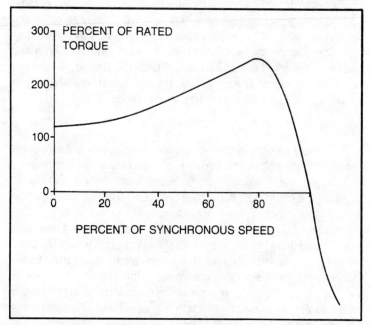

Fig. 2-6. A curve illustrating speed versus torque for an induction motor.

to the operating current of 0.82 amperes rms as the lamp lights. This starting surge is very brief, lasting only a few cycles; however, it is enough to make a small generator "hiccup," and the starting inrush must be considered if an electronic switch is used. A ratio of about 12:1 for inrush surge to operating current is not unusual for lamps and the lamp resistance actually changes during the cycle since the lamp tends to partially cool when the voltage is in the low part of the cycle. This accounts for the wave distortion seen in the oscillogram of Fig. 1-3.

Other heating elements such as irons and toasters also show an inrush surge. However, the ratio of inrush to operating current tends to be much lower—on the order of 2 to 4—the elements do not operate nearly as hot as a modern incandescent lamp filament. The inrush current tends to last much longer since the device usually takes several seconds to come to incandescence.

MAGNETIZATION

Items such as transformers (and to some extent induction motors), which operate by having the line current magnetize an iron core, are subject to another inrush phenomenon. If the core was last operated on say a positive-going cycle there will be some residual magnetism left in the core. If the next turn-on occurs on the same positive-going part of the wave, the core will usually saturate and the current will reach very high values limited mostly by the winding resistance during the first half cycle. This effect is very brief and must be considered mainly in terms of the half-cycle surge rating of electronic switches. The effect is worst in transformers which were operating without load when they were turned off.

RECTIFIERS

There are a certain number of electrical devices which operate using only one-half of the cycle. For example, some inexpensive speed-control drills will use a single silicon controlled rectifier (SCR) for control of the motor. Also some heat control devices will operate by supplying only half-wave current when the device is up to temperature and full-wave current when the device is below temperature. This yields a net DC term in the line current. The net DC term tends to bias the magnetism in the alternator or an inverter transformer away from zero and can cause saturation of the device in one-half cycle. This is particularly hard on small alternators and inverters and any condition in which a significant DC component develops should be avoided. These subjects will be discussed in greater length in the chapter dealing with power factor correction and load control.

GENERATORS AND ALTERNATORS

Under the right circumstances, nearly any machine which will operate as a motor can be made to operate as a generator, also. If the shaft of the device is simply turned faster than it would run as a motor, the counter emf will rise to exceed the line voltage and the machine will supply power to the line.

The machine of Fig. 2-5, for example, if supplied with a DC excitation on the rotor AB, will supply an alternating voltage at terminals CD and EF. If the rotor is turning 3600 rpm or 60 revolutions per second it will supply a 60-Hz voltage at these terminals. As noted before this is a two-phase machine in this configuration.

On the other hand, if the machine were to be excited with a DC current on windings CD and EF such that the fixed poles were magnetized N-S-N-S and the machine were rotated at 1800 rpm, the rotor winding AB would be excited with a 60-Hz voltage. In this case, the machine would be a single phase machine. As noted earlier, if the rotor were fitted with a commutator, the output of the brushes would be a pulsating DC. For a practical DC generator there would be a multiple set of windings on the rotor (or armature) to keep the pulsations down. On the other hand, if the current were taken off of the rotor with a pair of slip rings, this would be a practical alternating current machine.

The control of the output voltage would be obtained by means of the field excitation. At a fixed speed and load, the output voltage is a direct function of the field excitation up to the point where the iron begins to saturate. If the field current is doubled, the output voltage doubles. On most practical machines, the field saturation will begin at 10 to 20 percent above the rated voltage output of the machine.

As noted earlier, the armature reaction can have a substantial effect on this field, particularly in an AC machine. A very inductive load can make the output voltage sag badly. Since this is in the direction of reduced magnetism in the field, the voltage can be brought back by increasing the field current up to the point where the power dissipated in the field winding begins to overheat the machine. As a practical matter, this means that the machine will have only a limited ability to handle badly lagging loads.

In the capacitive load case, the tendency for the voltage to climb means that the field current must be reduced. It is not unusual to find that there is enough residual magnetism in the field for the machine to climb right up to saturation with no field current at all. Theoretically, it would be possible to supply a small amount of reverse excitation to partially cancel the residual magnetism in the field and bring the voltage back to normal. However, this is a delicate operation and probably not practical. As a practical matter it is usually better to prevent the machine from ever looking into a very large capacitive load.

In a home electrical system the existence of a large load with a leading phase angle would probably be very unlikely except for the fact that we will be introducing leading phase components for power factor correction. If the inductive load for which these components were inserted is suddenly removed, then the capacitive elements must be quickly removed in order to prevent the system voltage from soaring.

It is not widely appreciated that an induction motor can also be used for power generation and that it has some unique properties. If you will refer back to the curve of Fig. 2-6 you will see that at shaft speeds above the synchronous speed, the induction motor output torque is shown going negative. This means that the motor is absorbing torque from the shaft and that it is actually delivering power back into the line. Unlike the machine of Fig. 2-5, which must be operated at precisely 1800 or 3600 rpm in order to produce 60-Hz current, the induction machine will produce current precisely in step with the line over a range of speeds. Typically the machine would be running around rated load at a speed of about 110 percent of synchronous or $1.1 \times 1800 = 1980$ rpm. Useful output would be delivered from about 1900 to 2000 rpm, a variation of 5 percent.

In some cases, this speed tolerance is a very useful feature since it is not always possible to control the shaft speed of a prime mover such as a windmill or a water wheel to much better than this tolerance on a short term basis. A second advantage is the fact that the induction motor is the cheapest and simplest machine to acquire in a given rating.

The use of the induction motor in a generating capacity has one very substantial drawback. It can never be the only source of power in the system since it requires the presence of an alternating line voltage at its terminals *before* it can begin producing any electricity. This means that it can never be more than a helper in the system. A second and more subtle drawback is the fact that some rather special instrumentation is required to determine whether the machine is actually putting power into the system or extracting power from the system. Despite these factors, the ability to operate the induction machine as a helper without maintaining precise speed and phase synchronism with the remainder of the system is an advantage that bears consideration.

It should be noted that, once brought into synchronism with the line, a machine as shown in Fig. 2-5 will develop a powerful motor torque if it begins to lag and will assume most of the load and tend to slow if it begins to lead. Within limits, this action will tend to keep two machines synchronized once the initial synchronism is obtained. A wound rotor induction motor can usually be converted for generator usage.

Storage Batteries
and Solar Cells

It is nearly always necessary to have some form of electrical energy storage for a home electrical system to provide energy for the various operations which may be required when the main system is not operating. As a matter of fact there may be a requirement for some electrical energy to enable the main system to start up at all. You may need lamps for illumination and electricity to operate controls, pumps, and other items just to get the main system going. This seems like a good point in the book to get into a discussion of these systems since the storage battery-alternator-voltage regulator represents a simple self-sufficient system which is easily understood compared to some of the more sophisticated systems which will be required to operate the entire household.

Despite all of the effort which has gone into the development of alternate electrical energy storage techniques in recent years, the automotive lead-acid storage cell remains the most cost effective and viable mechanism for storing electrical energy. The economics of scale probably have a great deal to do with this. In a given year more than 10 million automotive storage batteries will be manufactured in the US. The nickle-cadmium cell and the silver-zinc cell and even the venerable iron-acid Edison cell may demonstrate advantages in terms of longer cycle life, increased energy storage density and sealed construction. However, if any significant storage capacity is contemplated, then the lead-acid cell comes out the clear winner in terms of reasonable first cost and perhaps life cycle cost. If a long life-cycle is contemplated, it is probably worthwhile to competitively price the Edison cell. However, in terms of ease of replacement and availability, the lead acid cell will probably still show an advantage. It is noteworthy that most of the manufacturers who provide standby power systems for computers make use of lead-acid storage.

A typical automobile battery selling for $40 will be rated at something like 60 ampere-hours and will generally be guaranteed to last four years in automotive use. The 60 ampere-hours implies 60 × 12V = 720 watt-hours of electrical energy. If we would presume that our battery bank would be required to provide a household load of 1.3 kW for 14 hours when the sun does not shine we would find that the system would require:

$$\frac{1,300 \text{ watts} \times 14 \text{ hours}}{720 \text{ watt-hours/battery}} = 25 \text{ batteries}$$

This represents an initial cost of $1000 in batteries alone. Assuming an inflation rate of eight percent, the next set of batteries will cost $1,361. Assuming a six percent return on investment (a savings reserve account) it will require something like $25 per month to provide the necessary reserve account to pay for the new set at the end of four years when they will presumably require replacement. Manipulated somewhat differently this works out to about $1 per month per 720 watt-hours of capacity.

From this calculation we can see that the provision for even a relatively modest amount of electrical energy storage would initially consume a significant fraction of the break-even investment and would exceed by a factor of three the $100 per year maintenance allowance. Allowing the same eight percent inflation rate previously used and the six percent return on investment, the figures work out to a battery replacement investment of $33.83 per month between the fourth and eighth years and $39.45 per month between the eighth and tenth years. This amounts to a total expenditure, including interest earnings of $5290.87. If we subtract this from the projected $7388 to pay for commercial electricity, we obtain $2097.13. This brings our initial break-even investment, exclusive of batteries, to $1171 when de-rated for 10 years of earnings at six percent. Viewed in a somewhat simpler manner we can see that the initial $25 per month represents a significant fraction of the initial power company rate of $39.35. Right at the start we see that we would have only $14.35 per month to work with if our goal is to save money on electrical energy.

Returning to some of the technical characteristics of the lead-acid cell we find that some of these are very well suited to the household energy storage problem. One of these is the fact that the lead-acid automotive battery is capable of sustaining tremendous overloads. When cranking an automobile engine, the battery is capable of supplying currents of 150 amperes when it is in good condition. It can also store the energy for periods of weeks or months with relatively little diminution of the stored quantity. Toward the end of the four-year life, these statements must be revised downward. However, we see that the battery bank is more than adequate for absorbing starting surges and performing the standby function for which it is intended.

On the negative side of the ledger, there are a number of substantial considerations. First and foremost is the fact that the electricity is stored in the wrong format. It is DC rather than AC. Connected in series, ten batteries can provide 112V DC. This source would be suitable for operation of incandescent lamps, certain power tools which have universal (AC or DC) brush motors and such. Unfortunately, it is not at all suitable for the operation of nearly any common household appliance. Connection of nearly any induction motor device such as a refrigerator, washer, sump-pump, automatic clothes washer or TV will bring about the rapid and dramatic demise of the device in question. Certain of these items, built for operation in yachts or campers, can be purchased for operation on 12V or 24V DC. However, they are generally small and very, very dear. A far more practical solution lies in the conversion of the DC to AC at 115V and 60 Hz so that ordinary home appliances can be made to operate from the prime storage medium. A subsequent chapter deals with this conversion.

The second drawback stems from the construction of the battery. The electrolyte used in these units is sulfuric acid. If the units are not sealed, the acid and acid fumes are continually vented to the atmosphere. This results in the rapid and destructive corrosion of terminals, fittings, racks and all manner of accoutrements surrounding the installation. In a fixed installation with a number of batteries, a fairly substantial amount of maintenance is required to wash down the batteries and neutralize the condensate with baking soda or some similar substance.

The third and perhaps most serious problem with a large battery bank is the fact that the batteries will generate a substantial amount of hydrogen gas under certain circumstances. Beneath the hood of an automobile, this gas is rapidly blown out to the atmosphere by the action of the engine fan and the wind. However, in a fixed battery bank, means must be provided for rapidly ventilating the area and preventing the pocketing of hydrogen under the ceiling.

A large battery bank is a definite no smoking area. Always provide plenty of ventilation. Never use brush motors or other spark producing devices in the battery bank area. Do not install a battery bank within the home.

THE GENERATOR

For many of the same reasons that the lead-acid cell is the logical choice for electrical storage, the automotive alternator represents a relatively logical choice for the prime generation of electricity to charge the batteries. To begin with, the alternator is made in such vast numbers that the economics of scale tend to reduce the price below that of nearly any comparable machine. Added to this is the fact that the auto graveyards gracing our countryside represent a vast repository of these machines, awaiting re-cycling. If the home-electrifier is so disposed and willing to remove the unit from the wrecked car himself, these units cost from $5 to $10 with perhaps the voltage regulator thrown in. If the unit is purchased as removed by the salvage operator, the price is likely to be closer to $25. The automotive alternator is too much of a bargain to be overlooked.

The automotive alternator is a truly flexible machine capable of satisfactory operation over a wide range of speed and of remarkably robust construction. It will survive the heat and vibration of the engine compartment and dousings with salt water with equanimity. A typical rating will be 14.2V DC output at 40 amperes or 568 watts. At this rating, it will absorb something on the order of 0.8 horsepower from the shaft. Three such machines are capable of providing our average 1.3 kW load and five or six, with the backup of the battery bank, would easily meet the normal peak/average ratios and keep the battery bank fully charged in the bargain. The machine takes full advantage of the economies of high frequency but outputs DC so that the problems of synchronization, phasing, and high frequency losses are eliminated. The illustration of Fig. 3-1 schematically depicts this machine.

In a typical automotive installation, the twelve-pole alternator rotates at twice the crankshaft speed and thus turns up something like 1200 rpm while idling and 6500 rpm at 60 mph. It is driven to speeds like 10,000 rpm at maximum engine speed. The view of the rotor assembly shows the rather unconventional shape of the pole pieces, and the armature assembly. The field winding is a straight spool of wire which is easy and inexpensive to wind. The pole pieces themselves are fluted cup-shaped affairs which are press-fitted together and almost completely encapsulate the field winding. This design has the advantage of providing the very high centrifugal bursting strength required for such a high-speed machine while at the same time lending itself to the extremely low-cost, automated assembly required of automobile components.

The three-phase stator is wired through a three-phase bridge rectifier which serves two purposes:

- •It rectifies the AC output of the stator for battery charging.
- •It prevents the battery from discharging back through the field and stator. This latter point eliminates the use of a high-current relay which used to be required.

The very wide speed range available with this machine can be a very marked advantage in home electrical installations since it makes possible the use of prime sources which have very poor speed regulation. For example, a windmill with a fixed pitch will tend to rotate at a speed proportional to wind velocity. Operation of a synchronous alternator would require some mechanism for keeping the speed of the alternator at precisely synchronous speed. With this machine the speed can be allowed to vary at well over more than an 8:1 range.

From the very earliest Kettering self-starter installations, it was recognized that automobile engines operate over widely vary-ing speed ranges and that the provision for an electrical mechanism for regulating the generator output voltage would be considerably cheaper and more reliable than a variable-speed transmission for the generator. The transistor/zener arrangement at the bottom of the illustration is a simplified representation of a modern solid-state voltage regulator.

The usual six-cell lead-acid battery in the car develops about 12.2 volts at rest and should not be continuously charged at voltages much higher than 14.2 volts. If the battery internal resistance is 0.1 ohm, the battery charging current would be approaching (4.2 - 12.2 volts) ÷ 0.1 ohm = 20 amperes at this level (actually the current will run a bit lower). However, since we have noted that the V_g is directly proportional to ω_μ then in a speedup from 12.3 volts at idling to top speed, the alternator output would rise by a factor of 10,000 ÷ 1200 = 8.33 to a level of 102.5 volts. Accordingly, it is necessary to provide some means of weakening the field by a factor of about eight in order to preserve the battery.

A simplified voltage regulator is shown in the schematic at the bottom of the figure. Diode VR_1 is arranged to go into conduction at about 12.5 volts. With the motor stopped or running slowly, the only current through R_3 is the base current of Q_1. Diode VR_2 serves to offset the voltage at the base of the transistor above point B and the transistor saturates with a current controlled by R_1 and the field drop plus the V_{sat} of the transistor. This provides maximum field to

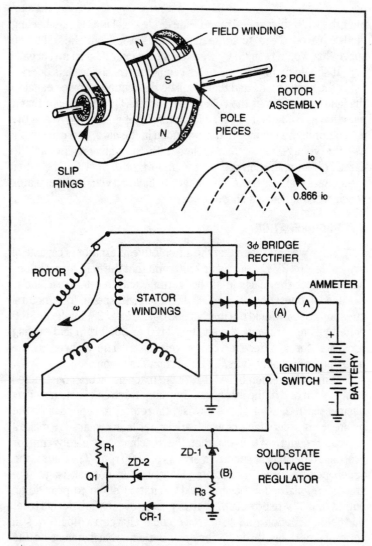

Fig. 3-1. The automotive alternator.

the alternator. As the alternator turns, the voltage at point A will rise and VR_1 will go into conduction which raises the voltage at B, thus reducing the forward bias of Q_1. The current at A is actually a humpy three-phase rectified affair; however, the voltage is largely smoothed by the battery. Nearly the entire voltage increment will appear at point B which will drastically lower the collector current of transistor Q_1. This in turn lowers the field current in the machine enough to control the output voltage of the machine to 14.2 volts over the wide shaft speed range. The diode CR-1 is a "free-wheeling diode" placed there to protect the transistor from the inductive surge that would result if Q_1 were suddenly cut off by a line surge.

It should be noted that in this regulator transistor Q_1 is operating in the linear mode and that it will be dissipating a power equal to the field current times the voltage drop across Q_1. This means that it must have an adequate heat sink. Naturally, if you intend to use the alternator in the manner that it was used in the auto, you can simply use the voltage regulator supplied with the unit initially and the details of the operation are not too important to you. On the other hand, there can be certain advantages to building your own regulator which we will discuss shortly.

HOT ROD OPERATION

Since we have seen that the automotive alternator is capable of providing a 14.2V DC output at 1200 rpm, that the field current does not overheat the machine at this setting, and that the machine is capable of operating at a much higher shaft speed, one natural question arises. Could the machine be operated at 28.4 volts or 36.6 volts by simply turning the shaft at 2400 or 3600 rpm and making provision for a different voltage regulator? The answer to the question is probably "yes," with certain limitations.

It must be remembered that these machines were constructed to be sold in one of the most fiercely competitive and cost conscious businesses in the world. It is unwise, therefore, to depend upon the existence of safety factors which are larger than required to make a reliable product. We know that the machine will safely run for protracted periods at speeds in excess of 7000 rpm. Therefore, the speed per se is not a problem. However, in normal usage, the voltage regulator knocks the field excitation down to provide an output of 14.2V to protect the battery. Suppose that we decide to let the field excitation rise so that we could charge two batteries in series or perhaps three batteries in series; what happens to the machine?

The first point to examine is the full-wave bridge rectifier. On most machines this consists of a series of diodes which are pressed into the stator structure for heat sinking. These diodes may or may not have an adequate peak-inverse voltage rating (PIV) to permit such operation. Some machines will use a GE model A44F or A44A diode in each of the six bridge positions. If these can be driven out and A44B or A44C diodes substituted satisfactorily, bridge operation can usually be assured at the higher voltage.

The second question concerns the turn-to-turn insulation and the turn-to-frame insulation. In general, this seems adequate for the higher voltage operation. However, an older machine with signs of heavy corrosion might be a good one to avoid. Careful cleaning and electrical varnishing of the coils can help this situation.

The third question concerns the shaft and pulley. Since the machine will be delivering two or three times as much output power, the shaft must accept two to three times as much input power as the unit does at 14.2V output. This means that the shaft must be able to withstand two or three times normal torque. The torque has to be delivered by the pulley. Actually this torque is no greater than the machine would have required when delivering 14.2V at 1200 rpm, but the combination of higher speed and torque could reduce shaft and bearing life. However, since shaft and bearing failures are not very common with these machines, the unit will probably operate satisfactorily from a mechanical viewpoint.

Because of the variety of alternators used on different cars, it is unwise to make too many generalizations. If the machine is bought new, the manufacturer may be queried about the operation under "hot rod" conditions. In the more likely case that the machine is purchased used, it can be tested. While I have actually operated a few machines at higher voltage without alteration, I would not recommend this for any application where reliability is of any concern. Probably the best bet would be to replace the bridge diodes and to clean and varnish the stator windings in any event. Even with this attention, there is little doubt that the higher voltage operation does not enhance the reliability of the machine.

SYSTEM CONNECTION

There is obviously something missing from the previous discussion. We have something like 25 batteries and three to six alternators but we have not discussed a mechanism for connecting these items together. An obvious mismatch exists between our generating capability and our storage capability. How do we best connect these items into an integrated system?

There are a very large number of ways in which the system connection could be accomplished. If we wished to have the facility of placing any battery in the system in any location, this could be accomplished with a crossbar switching matrix as sometimes used in telephone exchanges. However, this would require the use of some 1250 switches for the batteries alone. This is obviously not a very effective way of achieving minimum system cost. Therefore, let us examine some of the basic features that we might like to obtain from the system in order to provide maximum flexability at minimum system cost.

First of all, the system probably should provide something more than 12 to 14.2V output since the inverter will generally function more efficiently on a higher output. Secondly, the system should be capable of supplying some surge capability. However, this should be limited. If all of the batteries were simply wired in parallel, the system would have the capability of dumping something like 3750 amperes into a short circuit. A current of this magnitude will fuse a No. 1 B&S gauge wire in about eight milliseconds! This would be a pretty violent electrical explosion. Thirdly, it would be nice to be able to remove any battery from the system without having to shut down the system. There might be times when it was important to keep the electricity running and a "weak sister" was discovered in the bank in use.

The circuit shown schematically in Fig. 3-2 provides some of these features in reasonable measure. The batteries are organized into ranks and columns. At any one time only a single column may be connected across the string of alternators. When the string is to be changed, the second string is connected first and the first string is then removed. Probably a period of one to two hours per string should suffice to keep the entire bank charged as long as the alternators themselves are running.

It may be seen that with this arrangement, any of the busses may be grounded. Therefore any given battery need only have 12 to 14 volts with respect to ground. This voltage is low enough so that any battery may be removed from the circuit without danger while the bank is in operation, provided another column is in operation. The fifth alternator is shown as a standby unit. This may be used to keep the standby battery charged or to substitute for one of the four operational units so that the operational unit may be taken out of service for maintenance and such.

Figure 3-3 shows the suggested construction for the battery storage shelter. The use of cedar or cypress lumber in the battery area is recommended if a life in excess of 10 years is to be achieved. Similarly, we recommend stainless hardware. The battery room is constructed to take advantage of the siphon action due to solar heat and the prevailing wind. Fresh air enters through the screen at the bottom of the door and exits in the spaces between the roof rafters. Because of the differential velocity of the wind above the ground and the slope of the roof, the pressure at the top will be lower than the pressure at the bottom.

The switches in Fig. 3-2 may be replaced with heavy duty plug and jack arrangements. Alternatively, if automatic column switching is desired, relays may be used in these locations. Note that all switching and possible sparks take place on the switchboard side of the bulkhead.

Always make sure that a battery is disconnected from the circuit before attempting to remove it. Always leave the door open for a few minutes before entering the battery room. No smoking in the battery room.

SOLAR CELLS

As long as most of us can remember, there have been photosensitive cells which would provide a small amount of electricity when light fell upon them. Since the 1930's, photographers have used these lead-sulfide or lead selenide cells coupled to a very sensitive D'Arsonval meter to determine the amount of light in a scene for exposure setting. The amount of electricity produced by these devices was so small that they were of use only for measurement purposes.

As the techniques for grown-junction semiconductors began to develop, it became possible to exploit the phenomenon of the band gap. If photons with an energy greater than the band gap in a PN junction fall upon the junction, electron-hole pairs are formed. Pairs generated in and within a diffusion length of the junction can separate

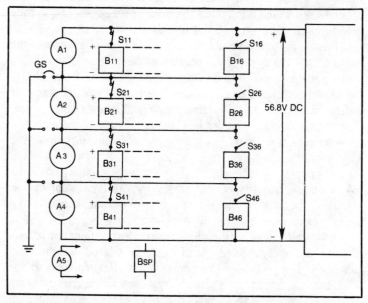

Fig. 3-2. An economical connection for 5 alternators and a total complement of 25 batteries is shown.

47

and contribute to an external current. If the junction is unbiased initially, it will become forward biased and a current will be produced in an external circuit. The device is then known as a photovoltaic cell, or if the unit is used to extract energy from sunlight it is called a solar cell. The efficiency of a solar cell depends upon the fraction of the sunlight that is reflected from the surface of the cell, the fraction absorbed in the transit to the junction and the fraction which separates the hole-electron pairs. Solar cells can be made from silicon or from gallium arsenide. On the basis of theoretical considerations the efficiency of the solar cell can reach 15 percent. However, practical solar cells tend to run in the three- to five-percent range.

The solar cell has proven to be a boon to the satellite business since it presents a nearly limitless quantity of electricity (if enough cells are used) and extracts the electricity directly from the sunshine with no moving parts, and no consumption of fuel or expendable materials. Nearly all of the satellites obtain their electrical power from solar cells. Even grade school pupils are familiar with the pictures of the broad solar panels on the communications satellites. From time to time we see articles in science magazines telling about the telephone company using solar cells for repeater power and read about solar cells being used to power all sorts of remote instruments. It really seems that these wondrous devices would be ideally suited to production of home electricity. Surely the achievable volume sales would bring down the price of solar cells just as they have brought down the price of transistors—or would they?

First of all, to answer that question let us consider some of the reasons why the price of the transistors have come down. When grown junction transistors first came upon the scene in the 1950's, most of the people making transistors were using 1.5-inch diameter wafers of silicon. On this wafer they could deposit through extremely fine masks something like 800 transistors. Of these, only 5 to 10 percent were good. As it was, the silicon used to make the wafers was the purest and most highly refined substance known to man, but the scattered impurities were frequent enough to spoil the overwhelming majority of the transistors. As time went by the refining process for the silicone improved and the masking techniques improved so that the transistors could be made smaller. The saws which sliced off the silicon wafers became finer so that less of the expensively refined silicon would be wasted and the yield increased. The processes have improved to the point now where three-inch diameter wafers of monocrystalline silicon are now being used by producers of LSI circuits and they can put the equivalent of about 100 million transistors on the wafer.

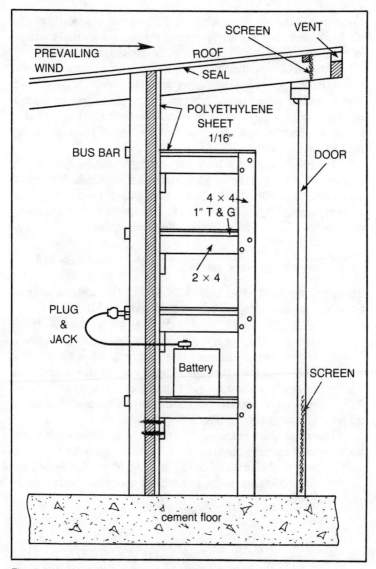

Fig. 3-3. In the battery rack area, lumber should be cedar or cypress for maximum life. Hardware should be stainless. Battery cables fit through tight holes in the polyethylene and plywood bulkhead.

That same wafer can be used to make *one* solar cell! For the solar cell the silicon does not have to be as pure as it does for transistors and there is not nearly the amount of labor involved. However, the point remains that as long as solar cells require the use of monocrystalline materials, the cell is going to be expensive.

Compared to the transistor whose size may be shrunk nearly indefinitely, the solar cell produces power on the basis of watts per square inch and one needs lots of square inches to get much power.

A potential breakthrough on the horizon seems imminent since certain laboratories have been experimenting with the production of polycrystalline solar cells which can be continuously cast from a machine in a long ribbon. These cells are less efficient but they may eventually become available at a tiny fraction of the cost of the monocrystalline cells.

The solar cell tends to have an operating voltage of about 0.5V, just about the same as the forward drop on a silicon diode. When well illuminated they will produce 0.14 to 0.16 amperes per square inch at a cost in small lots of $1.40 to $1.60 per square inch. If we are to consider our 1.3 kW average load we would obtain:

$$\frac{1,300 \text{ watts}}{0.5V \times 0.15 \text{ amperes per solar cell}} = 17,333 \text{ solar cells}$$

This array would cover an area of 120 square feet and at a price of $1.50 per square inch, the price would come to $26,000.

For a realistic system, the 1.3 kW does not tell the tale. If the sun shone only a third of the time on the cells, and the battery storage system had an efficiency of 33 percent, we would actually require a 9.3-kW array to both provide power and charge batteries during the eight hours of sunshine. In this case, the array amounts to 122,384 solar cells. The array would measure 850 square feet (29 × 29 feet) and would cost $183,600!

The point does not deserve further belaboring. Until some breakthrough reduces the cost of solar cells to a tiny fraction (like one percent of the present cost), the use of solar cells will be confined to multi-million dollar satellites or very small, very expensive installations.

DC to AC Conversion

It was noted earlier that for all practical purposes it is necessary to convert the stored DC energy in the battery bank into alternating current if something other than a few lamps and perhaps some hand tools are to be operated during periods when the prime power source is idle because the sun is not shining or the wind is not blowing. In this chapter we will attempt to describe some of the ways in which this DC to AC conversion may be accomplished.

FREQUENCY AND WAVESHAPE CONTROL

To begin with let us discuss some of the requirements of your home electrical system in terms of the frequency and the waveshape. The power company controls the frequency of the power supplied to you within rather close tolerances. In general, the frequency measured on the power line does not depart by more than ± 0.2 Hz measured over short periods such as a half cycle and the stability averaged over a day is usually within one part in 10^6. The reason for this is the fact that the systems are intertied. The Rochester Gas and Electric company will sometimes buy power from Ontario Hydro, Con Edison or Niagara Mohawk through the New York State Power Pool. A pair of machines such as shown in Fig. 2-5 may be tied together if they are operating at the same speed, and at nearly the same phase angle. If we presume that the DC excitation is applied to the rotor windings AB and the output taken from the field windings CD and EF, two such machines may be paralleled by connecting E_1 and E_2 and F_1 and F_2 . . . etc., provided that the instantaneous voltages EF_1 and EF_2 and such are essentially identical. Under these conditions we would find that a torque curve such as the one shown in Fig. 4-1 would be obtained for either of the machines. The torque required to turn the shaft of either machine would depend upon the angular difference between the shafts of the two machines and the field excitation. If the shaft of one machine were to begin to lag behind, it will begin to act as a motor and extract power from the line. This would reduce the torque required from its prime mover which would creep up and reduce the angle. Application of more torque to the shaft of either machine would increase the fraction of the system load borne by that machine. Within the range bounded by the pull-out torque and the force-out torque, the machines will be synchronously locked and can share the external load with the work division determined by the amount of torque applied and the excitation. If synchronism is lost, the machines will fight one another disastrously and must be immediately disconnected.

Obviously, it is much easier to provide for synchronism in a large system having many contributing machines if the frequency is agreed upon and rigorously adhered to. However, in a small system for home electricity, is it necessary to maintain such precise control? The answer is that it may be to your advantage to maintain precise frequency control for a number of reasons.

First of all, it may be of advantage to operate two machines in synchronism in a load sharing manner for a number of reasons. For example, a small hydro installation might be used for standby power to back up a wind installation. When the wind is blowing strongly, the hydro could be shut down entirely and in periods of light wind, the hydro might share the load with the wind installation. This would conserve water in the impoundment which is usually a big consideration in hydro installations.

Secondly, because of the fact that the power frequency is known to be precisely controlled, many of the appliances in the home employ the power frequency for timing purposes. The timers on your kitchen range, the microwave oven, the automatic clothes washer, and dishwasher all operate on the power frequency. The picture on your TV set is synchronized to the power frequency so that interference from the power line is minimized and the rotational speed of all but the most expensive turntables is determined by the power frequency. In addition to this, nearly all of the electric clocks about the home keep time by synchronizing with the power frequency. If the frequency were allowed to err by 1 Hz, the clock would gain or lose 1 minute per hour or 24 minutes per day.

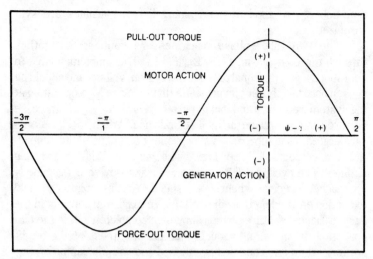

Fig. 4-1. The synchronous motor torque curve.

Therefore, it may be seen that a great many of the uses made of electricity in the home require a very precise control of frequency. In some cases such as the microwave oven, the error of one part in sixty in timing may not be critical. However, in others such as a stereo, the records and tapes will play sharp or flat by noticeable amounts. The TV may be afflicted with fluttering or creeping ghosts which can be rather annoying to view.

Though less critical, another reason for maintaining the power frequency close to 60 Hz stems from the fact that with increasing frequency a capacitor will draw more current whereas an inductor will draw less. Circuits such as power factor correction and filtering circuits rapidly unbalance as a function of power frequency. There are a great number of such circuits hidden in modern appliances. Unless the appliance carries the legend 115V/50-65 Hz or something similar, it is difficult to guarantee that the appliance will function on "off-frequency" power.

For these reasons, we recommend a precision frequency control will yield the most effective and satisfactory home electrical system. A more sloppy system will do some of the things you have learned to expect from electrical power but not all. Techniques for obtaining high precision 60-Hz power will be discussed shortly.

SYNCHRONIZING MACHINES

Before leaving this point, it is worthwhile to examine a simple technique for synchronizing two alternators. Figure 4-2 shows a time-honored method of determining if two machines are synchronized.

The two three-phase machines are connected together through three large lamps, L_1, L_2, and L_3. The lamps may have to operate on a voltage equal to twice the system voltage and should be chosen accordingly. Let us presume that machine M is operating at the system frequency and that generator N is stationary. In this case all three lamps will be lit at half brilliance. When machine N is accelerated, as it approaches synchronous speed, if all the connections are correct, all three lamps will begin to go brighter and then dimmer. The cycling will become slower and slower as machine N approaches precise synchronous speed. At this point, the field excitation on winding IJ is adjusted to give maximum contrast in the lamp brilliance. All three lamps should go completely out in the dim portion of the cycle and should reach full brightness on the bright portion. When this condition is obtained and with the lamps cycling at something slower than once per second, switch S_1 may be closed the next time all three lamps are out. The synchronizing torque illustrated in Fig. 4-1 will then tend to hold the two machines in synchronism and the split of the load will be controlled by the field excitation and the shaft torque of each machine.

Sometimes on a first connection it will be found possible to make one of the lamps, say L_1, go out while L_2 and L_3 are burning brightly. In this case the phase rotation of the two machines is reversed. This can be remedied by swapping either connections E and F or B and C, or by reversing the direction of shaft rotation on one of the machines. Reversal of one of the winding terminal pairs is usually easier than reversing the shaft rotation.

The use of the lamps to detect the difference in voltage between the generators on an instantaneous basis works equally well on single phase or split phase systems where one or two lamps

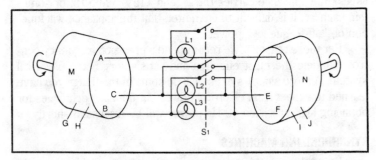

Fig. 4-2. A time-honored method of determining if two machines are synchronized.

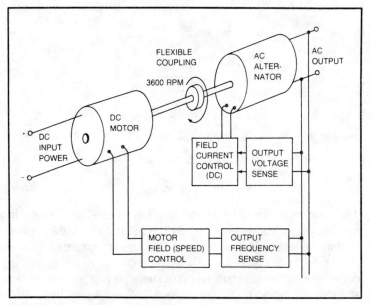

Fig. 4-3. The motor-generator set.

would be placed in the hot leads. Never interrupt the ground lead. Fuses have been left out of the diagram for simplicity

Do not leave out the fuses in an actual setup.

WAVESHAPE

As seen in the earlier oscillograms, the voltage supplied to you by the power company takes the form of sinewaves at a frequency of 60 Hz. It is usually rated at 115V rms or 110V rms which means that it will develop as much heat in a given resistor as 115V DC or 110V DC. The sine wave voltage actually follows the relationship:

$$E = 162.6V \sin (360° \times 60 \times t)$$

for 115V rms

where t = time in seconds.

This says that the voltage actually gets as high as ± 162.6V for brief periods in the cycle and is lower the rest of the time.

Now suppose that we were to simply take a string of batteries which added up to 162.6 volts and arrange a switch to reverse the connection between the battery and the load 120 times per second. This would be a 60-Hz squarewave. If this were connected to a load consisting of a 100-ohm resistor, the peak current would be equal to the peak current for the 115V AC source. However, the power dissipated in the resistor would be:

$$\text{Power} = \frac{E^2}{R}$$

$$= \frac{(162.6)^2}{100}$$

$$= 246.5 \text{ watts}$$

For the 115V rms case we would have

$$\text{Power} = \frac{(115)^2}{100}$$

$$= 132.25 \text{ watts}$$

The squarewave would be dissipating 1.414 times as much power in the resistor. This tells us that the corners of the square wave contain an extra 41.4 percent of the power compared to the sinewave. Since the torque of a typical induction motor is largely determined by the peak current of the sinewave component, we can see that we would be putting a good deal extra into the motor without getting much extra out. This is somewhat of an oversimplification, but it serves to illustrate the point. In general, the corners of the squarewave can be shown to consist of higher frequency components, odd multiples of the 60-Hz frequency. Because the laminations of the typical motor and transformer are too thick, frequencies above 60 Hz tend to be dissipated in eddy currents flowing in the laminations. The cores can become very hot, frequently hot enough to burn out the component.

If we were to reduce the battery voltage to 115V DC, we would make the power dissipated in the resistor equal to the AC case. However, the output voltage of a transformer-rectifier combination as used in a stereo or TV would be only 71 percent of normal and the torque of an induction motor would be only about 60 percent as great. Furthermore, both units would still experience excessive heating. Lampbulbs and some universal handtool motors would fare better.

In order to successfully operate the full spectrum of home appliances, it is necessary to provide a good sine wave output. This is a point of particular concern if one of the low-cost inverters sold at radio stores is considered. In order to operate with minimum iron in the transformer (thereby saving cost and weight) these units generally produce high-frequency square waves. These units will generally not operate anything with an induction motor and will in fact usually destroy themselves or the appliance if left connected for very long.

THE MOTOR GENERATOR SET

Perhaps the easiest and most straightforward way to convert DC to AC and produce good sine wave output is to use a DC motor to turn an AC alternator. If the motor speed is controlled to turn the alternator at precisely synchronous speed, the system can very nicely provide 60-Hz sine waves of good waveform. The qualifier "good," rather than "perfect," is because it is difficult to design small machines to provide an exact sine wave output. However, a well-designed machine can be built to have no more than 1- or 2-percent harmonic content.

Motor generator sets can be purchased in a wide variety of styles and ratings, including machines that are packaged in a single frame with no shaft output. However, for purposes of generality, the arrangement using a separate motor and generator with the shafts tied together through a flexible coupling is shown in Fig. 4-3.

In this arrangement a DC motor provides the turning shaft to power the AC alternator. In this case the alternator is assumed to be a two-pole 3600 rpm machine. Hanging across the alternator output is a device marked *output voltage sense*. This device senses the output voltage and drives a field current control for the alternator. If the output voltage of the alternator sags due to a heavy resistive load or an inductive load, the field current control increases the field current enough to bring the output voltage back to the desired 115V rms.

The second sensor across the output line senses the output frequency of the alternator. If this frequency is low, the motor speed control *reduces* the field to the motor which makes the motor draw more current and run faster. This point bears a little explanation. The current drawn by the DC motor is a function of the difference between the line voltage and the motor counter emf and the resistance of the motor windings. When the field is reduced in the motor, the counter emf falls and the motor draws more current which causes it to pick up speed to find a new equilibrium point. Conversely, raising the motor field increases the counter emf and causes the motor to draw less current and to slow down.

While it is not possible in a text of this size and covering the breadth of subject matter which must be treated to provide a definitive treatment of items like motor and generator regulators and controls, it is perhaps instructive to at least briefly look at some of these items to obtain a feel for the operating principles.

Figure 4-4 shows a simple voltage control for an alternator. An AC line transformer develops an isolated voltage proportional to the

AC line voltage. This is rectified in a full-wave rectifier and applied to a three-terminal voltage regulator. These devices will develop a voltage which is nearly independent of load and temperature. The op amp compares a sample of the rectified signal with a 7.5V reference derived from the regulator output. If the rectified sample exceeds the 7.5 volts, the output voltage of the op amp falls, reducing the current through the Darlington pair Q_1 and Q_2. This in turn reduces the field in the alternator and the alternator output voltage. The voltage control may be used to set the level of the peak AC line voltage.

The gain control serves a somewhat different function. Because this is a full closed-loop servo system, it is entirely possible for the system to swing into a low-frequency oscillation. The gain control should be operated at a gain level where this does not happen.

Diode CR_1 is employed to give the alternator magnetic field a "free-wheeling" path in which to collapse without destroying the pass transistors Q_1 and Q_2 in the event that they become cutoff. CR_1 is normally back-biased. Note that the pass transistors may be operating in the linear mode and thereby dissipating a good deal of power.

The voltage at A actually has a certain amount of ripple caused by the charging to replenish the rectifier filter. If the gain of the op amp is set high enough, this can actually force the unit into a switching mode operation and the collector voltage of Q_1 will be a rectangular wave at 120 Hz. Operating in this mode, the control will be duty-factor modulated and the dissipation in the pass transistors will be much smaller. This may or may not work well depending upon the machine. To eliminate this action, the gain control may be shunted with a capacitor to reduce the gain of the op amp at 120 Hz. A capacitor at the sample tap point will also perform the same function.

THE SPEED OR FREQUENCY CONTROL

To begin with, let us consider the properties of the driving DC motor to obtain a feel for the amount of control effort that is required. We shall consider the possibility of controlling both the field of the motor and the armature with the same machine. As noted earlier, if the output load of the machine is increased, the machine will want to slow down. To prevent the slowing, which would change the operating frequency, the motor must draw more current. The current that the motor will draw is given by:

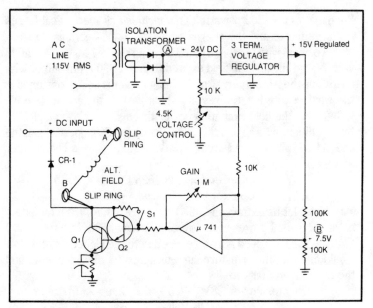

Fig. 4-4. A solid-state voltage control.

Line Voltage = counter emf + (armature resistance × armature current)

The counter ems is the counter electromotive force generated by the motor.

If we look at this equation, we can see that the current can be raised by either lowering the counter emf or raising the line voltage. The counter emf is a direct function of the rotational speed of the armature and the field strength. If we do nothing in the way of control, the motor will simply slow until the counter emf falls to the point where the motor is drawing enough current to come to a new equilibrium. However, this is just what we want to prevent. But if we weaken the field by the correct amount by reducing the field current, the counter emf can fall without any slowing of the motor, and the motor will draw enough current to provide the torque for the new increased load.

For our example, let us consider a system with the following parameters:

Motor speed = 3600 rpm

Nominal line voltage = 28V DC

Armature resistance, Low loss = .004628 ohms

High Loss = .0127 ohms

Full output load = 3500 watts (resistive load)

Windage loss = 262.5 watts (This includes all mechanical losses from the motor and the alternator windage and bearings, etc.)

To begin with, suppose that we set up the low-loss motor at 28 volts on the line and the field excited to give a 3600 rpm output with no load on the AC output; all of the input power is being consumed in the windage loss and the resistive heat loss on the armature. (We will neglect the field excitation power throughout this example.) In this condition, we find that the unit is drawing 9.39 amperes from the line. This is an input of 9.39 A × 28V = 262.9 watts. The counter emf is:

$$28V = \text{counter emf} + (.004628 \text{ Ohms} \times 9.39)$$
$$\text{Counter emf} = 27.96 \text{ Volts}$$

Now, let us extract the full 3.5 kW (resistive) from the alternator. The total power in the system is now 3500 Watts + 262.9 Watts + the increased armature loss due to the higher current. We would find that the armature current jumps to 137.5 amperes and the counter emf falls to

$$\text{Counter emf} = 28V - (.004628 \text{ Ohms} \times 137.5)$$
$$= 27.36 \text{ Volts}$$

Because the speed remains constant, the windage loss remains constant at 262.5 watts, the output is now 3500 Watts and the I^2R loss in the armature is the difference between the 3850-watt input and the 3762.5-watt output plus windage, or 87.5 watts.

In order to hold the speed constant, we have to reduce the field excitation by the ratio 27.96/27.36 = 102.19 percent. Without the control, the speed would have sagged by nearly the same ratio, and the frequency would have fallen to 58.7 Hz.

In the high-loss case where the armature resistance is .0127 ohms, we would find that the no-load counter emf is 27.88V and the 3.5 kW output counter emf is 26.17V. Therefore, it would be necessary to weaken the field by a factor of 27.88/26.17, or 1.065. The loaded armature current would be 143.75 amperes in this case, and the I^2R loss in the armature would be 262.5 watts.

ARMATURE CONTROL

In the case of the armature control running with full load plus armature and windage losses we find that the figures work out the same as they did for the field-weakening example. However, when the load is removed from the alternator, it becomes necessary to reduce the line voltage to prevent the unit from overspeeding. Since neither the field nor the speed are changing in this case, the counter emf remains constant at 27.36 volts and the line current is 9.38 amperes. The necessary line voltage becomes: (low loss case)

Line Voltage = 27.36 + (0.004628 ohms × 9.38 A)
          = 27.40 Volts

In the high-loss case, it would have been necessary to reduce the line voltage to 26.29 volts (Note the field would have been set weaker in the high loss case to hold the full load speed at 3600 rpm. This accounts for the seeming inconsistency of the numbers.) The line voltage ratios are 1.02 and 1.065, respectively, just as in the field weakening example.

There are several conclusions to be drawn from the example.

- There is no net advantage to either field-weakening or armature control from an efficiency or a sensitivity standpoint.
- The amount of control required is directly proportional to armature resistance. A low resistance machine is very sensitive to control.
- Without some control the output frequency would vary significantly. 2 percent in the low-loss case and 6.5 percent in the high-loss case.

There is a net advantage in the field control case due to the fact that the field current is very much smaller than the armature current. This means that the transistors used to control are much smaller and would run a great deal cooler. For a machine of this size, a field current less than an ampere would not be unusual, whereas the control of the armature would require transistors capable of handling currents up to 145 amperes in the high-loss case.

A second point of interest is the resistance values. In the high-loss case, we see that the machine would be dissipating 262 watts of I^2R loss in the armature and some fraction of the total 262.5 Watts windage loss in the motor. This is a good deal of heat and would make the motor quite hot in continuous operation. In terms of power loss, the two cases represent 10 and 15 percent efficiency so that the overall performance is not greatly affected. But the heating is another matter. Most motors are rated for power by the amount of heat they can dissipate, without having the machine temperature rise more than 40°C. Depending upon the effectiveness of the fan, the high-loss motor might not be capable of operating at a continuous load of 3.5 kW. In selection of a motor, it is advisable to get a low resistance unit.

A method of obtaining the temperature rise of the motor is to measure the resistance of the winding at room temperature and to measure the resistance with the motor hot. The resistance of hard-drawn copper wire rises by a factor of .00382 per °C. Thus, the resistance of the armature should change by (.00382 × 40) +1 = 1.153 for a 40°C rise.

The examples selected are representative of fairly high quality machines. Certain factors such as brush drop have been neglected, but the results are reasonably close to the facts. The results would be similar for motors of other voltage and power ratings.

SPEED SENSING

As we have seen from the examples, the regulation changes are so small in a reasonably efficient system—on the order of 2 to 6.5 percent—that the use of an open-loop speed correction system is probably not practical. The variations due to temperature and other factors such as changing brush drop require that a reasonably regulated system must have a closed loop sensing system. We shall next deal with two classes of speed sensors.

THE ONE-SHOT TACHOMETER

There is a large class of devices which I shall lump together in the class of one-shot tachometers. In essence all of these devices operate on the principle of providing a fixed quantum of energy into an integrating circuit for each increment of angular rotation. The integrated output is then a function of rotational velocity (or other parameter). Figure 4-5 illustrates this device in its simplest form. The battery charges the capacitor up to a fixed level every time the switch closes in that direction and then discharges it through the meter in the opposite position. The average current through the meter is proportional to the number of closures per second.

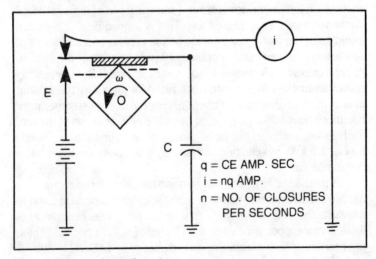

Fig. 4-5. The one-shot tachometer.

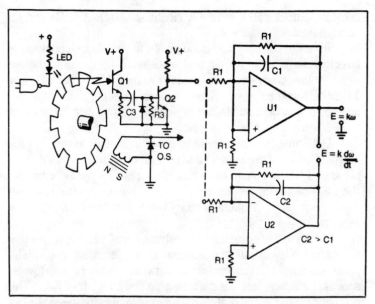

Fig. 4-6. The optical tachometer and the magnetic tachometer.

The circuits of Fig. 4-6 illustrate a variation of this scheme in which a slotted disc is used as a photon interruptor. It breaks the light path between the LED and Q_1. When Q_1 is illuminated, it saturates and delivers a short spike to the base of Q_2—through C_3 and R_3, which have been chosen to have a time constant much shorter than the shortest illumination period. Amplifier U_1 is arranged as an integrator, or low-pass filter, to average out the charge transfer. This is not a true integrator in the mathematical sense but rather a low-pass averaging filter whose time constant is determined by the values of R_1 and C_1. If a second such active filter with a longer time constant is used as a reference, the angular acceleration of the unit is obtainable from the difference between the output terminals. In the configuration shown, the output of U_1 would run between $-(V+)$ at zero velocity and nearly $-(V+/2)$ at maximum speed when C_3R_3 was nearly equal to the on time of Q_1. Up to nearly the maximum limit, the unit is quite linear in output.

Also shown in the figure is a magnetic pickup made up of a bar magnet with a winding. The zener diode will transform the output pulse into a nearly square wave above some minimal speed by clamping the negative swing to nearly ground and clipping the positive swing. This action limits the output voltage to an essentially constant voltage wave, but the duty cycle is also constant. It is therefore necessary to drive a one-shot of some form to obtain a constant output pulse width. A circuit similar to the Q_1 and Q_2 arrangement will suffice.

Since the accuracy and linearity of the one-shot tachometer is directly related to the stability and precision of the one-shot, the use of one of the IC one-shot devices such as the N555 or the 74121 or 74122 is recommended. These devices have been carefully designed to provide a stable and repeatable output with variation in temperature, supply voltage, etc.

The Photo Interrupter can be either a made up assembly such as the General Electric H-11 series or may be fabricated from separate LED and photo transistor elements. It is noteworthy that the LED can be driven by TTL logic as shown, and the one-shot mechanism can be implemented by turning off the LED after the initial pulse.

Both types of units have the advantage of offering no friction load although the magnetic pickup may have a detenting torque. The photo transistor arrangement is workable down to the lowest speeds, including zero but must be shielded from stray light. The magnetic pickup is inoperative below some minimum speed but can often be easily implemented by placing the pickup near an existing iron or steel (but not stainless) gear. An earphone magnet makes a very good pickup. However suitable magnetic pickups are available commercially. Hall effect magnetic pickups, which are workable down to zero velocity, are also available. These units often feature direct TTL output.

These tachometer techniques are essentially analog in nature. They can be made to be stable and repeatable to better than one percent with a little care and can be improved to 0.1 percent with a great deal of care. This order of accuracy requires a very careful compensation of temperature variation and such in order to minimize drift.

If the system goal does not include the operation of clocks or very fine phase matching, then this form of speed sensor is probably satisfactory. However, for any operations in which clocks and timers are to be used, the employment of a phase-locked loop is probably imperative.

PHASE LOCKED LOOP CONTROLS

When a really precise speed control is required, the choice usually goes to a phase locked loop (PLL). There are a variety of reasons for this choice but the principal one is the fact that there are few things in this world which can be measured with the accuracy that can be obtained in frequency and time measurements. A rather modest quartz watch will keep time to within about a minute per month, which represents a speed error of about one part in 43,200 or 2.3×10^{-5}. With a little bit of care, a nonovened 5-MHz quartz oscillator can be synchronized with National Bureau of Standards WWV radio transmission to one part in 10^6 or better. With the addition of an oven, or other temperature control, a stability of one in 10^8 is obtainable. If you had a scales this accurate, you could take two successive weighings of the aircraft carrier Enterprise and tell whether the Captain had carried his sextant aboard! If you want to get really precise you can purchase a cesium or rubidium standard with a guaranteed accuracy over the long term of ± 3 in 10^{-12} which is settable to one in 10^{13}. For the long term rate, a cesium clock set at the time of the birth of Christ could by now have accumulated an error of 0.0062 seconds (if it hadn't given up long ago). And one of the best parts of this is the fact that the accuracy of frequency standards can be verified against the National Bureau of Standards by radio, anywhere in the world, using equipment available to the most modest laboratories.

In order to operate a phase-locked loop it is necessary that we have some mechanism for detecting the relative phase of two waves. We shall begin with the sine wave phase detector.

THE SINE WAVE PHASE DETECTOR

The circuit shown in Fig. 4-7 has been around for a long time. It was originally used by Major Armstrong as the detector in an FM radio. We shall not treat the design of this circuit since it can be found in nearly any radio handbook. However, it is worthwhile to have a look at how it operates.

The operation of the transformer T1 is such that it drives points L and M out of phase whereas T2 feeds L and M in-phase. In Fig. 4-8

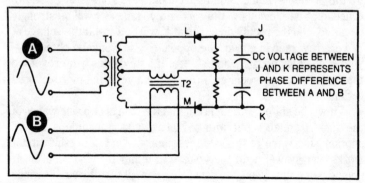

Fig. 4-7. The sine wave phase detector.

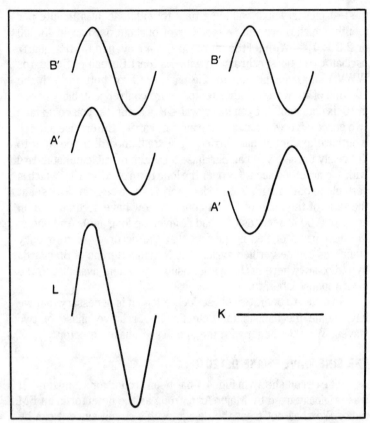

Fig. 4-8. Voltage addition at point L of Fig. 4-7 when inputs A and B are in-phase.

the special case for inputs A and B exactly in-phase and equal in amplitude is shown. The induced components, which are designated by the primed symbols, add up to double the voltage at L, and to cancel it at K. Obviously, the rectified voltage at J will be strongly negative, and the rectified voltage at K will be zero. If the voltage at B were reversed in phase, the opposite case would be obtained—with K strongly negative and L zero. If both diodes were reversed the output voltage would go strongly positive at the more favored terminal.

It will be stated without proof that the sum of two sine waves of the same frequency but with any phase and amplitude is always another sine wave of the same frequency. Therefore, this circuit works equally well with the voltage at either terminal smaller. It should be relatively easy to convince yourself that the circuit winds up with the rectified output at J and K equal when the inputs are in quadrature or have a 90° phase difference. The circuit will also work well with any waveform which is symmetrical and has equal positive and negative halves.

The latter feature is particularly handy in view of the fact that it is much easier to derive a square wave from a counter chain than it is to derive a sine wave. We can take any quartz crystal, count the output down to a suitable standard frequency and use that for the reference applied to terminal B.

Terminal A could be hooked across the 60-Hz output of our alternator, and the voltage difference between terminals J and K will reflect the phase difference between our quartz-controlled reference and the alternator voltage.

It should be noted that the two signals to be compared should be symmetrical; that is, they must have equal positive and negative halves. This can be assured if the divisor in the counter chain always divides by an even number.

THE DIGITAL PHASE DETECTOR

The circuit of Fig. 4-7 requires the use of a pair of transformers, and the circuit is somewhat analog in nature. While it will work

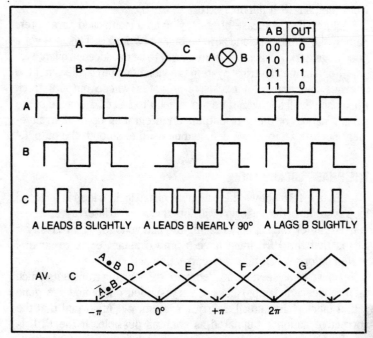

Fig. 4-9. The EXCLUSIVE OR digital phase detector.

with digital signals isn't there a way to compare the phase of digital signals and for that matter, what is the meaning of the phase of an on and off digital signal? Let us suppose that the signals at A and B are both perfect square waves (implying a 50-percent duty cycle). Figure 4-9 shows the use of an EXCLUSIVE OR gate as a phase comparator. As noted from the truth table, this unit gives a true, or high, output only when the signals are different and not when they are alike. From this you can see that when A and B are perfectly in phase the output of the gate will be always zero, and when A and B are exactly opposite, or 180° out of phase, the output will always be one. For points in between, the unit tends to produce a pulse output whose average value is a function of the phase angle. The average value of output C is shown in the accompanying graph.

The output of a typical logic gate is typically not voltage controlled with any great precision, so the use of a voltage average of the output waveform is not a particularly accurate reference. However, EXCLUSIVE OR gates usually come more than one to a package and they *generally* will track one another pretty well with temperature. In this case if A is inverted, it will be seen that the output of the second EXCLUSIVE OR will tend to have the opposite slope. In this case a high CMRR amplifier will be able to derive the sense signal from the pair with minimal error.

Note: This is a case of employing an uncontrolled parameter, namely the output one and zero voltages of a device. This is never a good engineering practice for any circuit to be produced in quantity.

For a home electrical system the violation of this rule is not too onerous and a pair of exclusive-or gates whose outputs track reasonably well should not be too difficult to select. Because of the pulsing nature of the gate outputs an integrating op-amp arrangement such as U1 in Fig. 4-6 should be used to smooth the output.

THE PHASE LOCKED LOOP

Figure 4-10 shows the complete phase locked loop in a closed loop system. In this case you will note that a photo interrupter wheel is used rather than simply reading the line voltage. The reason for this is that the PLL must have a time constant equal to several waves before it can take any corrective action.

As we have seen earlier, some of the starting transients (such as incandescent lamps) have a very sudden onset and are gone rather quickly. In a small electrical system, it is important that the corrective action be initiated as soon as possible. If the PLL is synchronizing to some frequency much higher than the 60-Hz out-

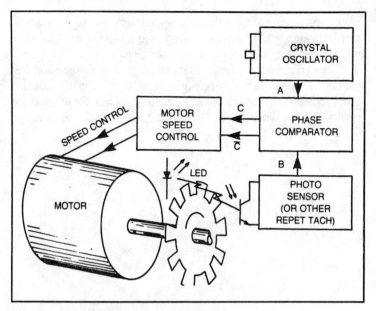

Fig. 4-10. A phase locked loop speed control.

put of the alternator, this rapid response is possible in a fraction of a revolution.

The photo interruptor is compared for phase with the reference signal derived from the crystal oscillator and counter chain. If the motor starts to get ahead, a signal is sent to the speed control, which calls for a slower speed, probably by reducing the voltage or increasing field current. If the motor starts to lag, the converse correction is applied.

The comparison between this form of system and the one-shot tachometer control is about the same as the comparison between a Wheatstone bridge and an ohmmeter. In the one-shot system, any variation in the capacitance or the period of the one-shot will give a corresponding shift in the motor speed. Similarly, any drift in the zero setting will be reflected as a change in motor speed. These are probably not large errors, but they can be eliminated in the phase locked loop system. Here, the interruptor phase is continuously being compared with the phase of the crystal derived frequency. The system has its own kinds of errors, but they can be made to be orders of magnitude smaller than the one-shot system errors.

The reduction in the time lag facilitated by operating at a multiple of the 60-Hz frequency permits the system to operate more smoothly since the corrective action will be initiated before the motor will have had time to lag in speed very much. A magnetic sensor using a gear can also be used, and a veneer saw blade can make a fine precision photo interruptor disc.

The treatment of the problem of controlling voltage and frequency has been made necessarily brief and cursory in this chapter. For a somewhat broader and more detailed look into these areas the reader is again directed to: TAB book No. 929, *Solid-State Motor Controls*.

5

The Inverter

Compared to the motor generator set, the solid-state inverter would seem to hold a great deal of appeal. It has no rotating parts which wear and require lubrication. It has no brushes to wear out and require replacement and generate dirt. It has no commutator which requires periodic turning and undercutting. It furthermore should be smaller, lighter, and more convenient to use—and perhaps in the long run—it should be less expensive.

An additional line of appeal is the fact that the inverter does not require precision machine tools to build. It can be fabricated in the typical home workshop very easily. A good many radio hams have fabricated inverter supplies for their radios to permit mobile or portable operation.

With all of these advantages it perhaps comes as a surprise to find the previous chapter devoted to motor generator sets and the precise control and synchronization of same. Let's take a look at the inverter and see whether we can explain some of the reasons why the motor-generator set manages to survive in the solid-state age.

To begin with let us examine the free-running or self-controlled inverter of Fig. 5-1. This is similar to the devices one finds for sale in a radio store which will plug into the cigarette lighter in your car to power 110-volt lamps, shavers, or other light appliances. The voltage and current waveforms are shown below to assist in the explanation.

The two transistors Q_1 and Q_2 are initially biased on slightly. One or the other of these will draw slightly more current than the

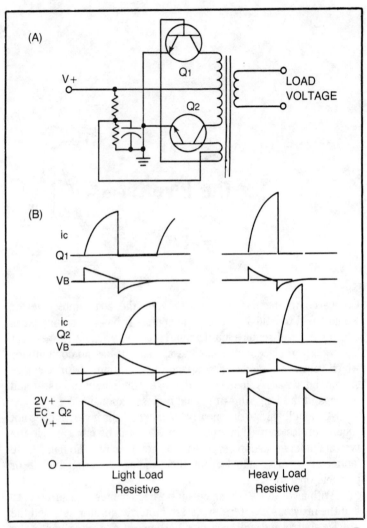

Fig. 5-1. The free-running inverter at A and voltage and current waveforms at B.

other, and the voltage induced in the feedback winding will forward bias this unit and reverse bias the other until the first unit is saturated and the other is cutoff. The collector current will begin to climb toward the level established by the IR drop of the winding of the ON section. As the time rate of climb of this current falls off and the current begins to knee over, the forward bias falls and the reverse bias of the other transistor diminishes until a point is reached where the second unit begins to conduct. This rapidly switches the first unit OFF and the second unit ON. The cycle repeats for the second unit. During the OFF cycle, the cutoff transistor experiences a collector voltage equal to twice the supply voltage. The output voltage tends to be the same waveshape as V_b; that is, a sagging square wave.

In the case of heavier loads, the current rises faster and the switching happens more frequently; therefore, the output frequency is higher. If the load resistance is too low and the load wants to draw too much current, the feedback voltage can fall to the point where one of the transistors will no longer turn off. In this condition, with both Q_1 and Q_2 biased ON, the unit will simply draw a large enough current to destroy one or both of the transistors unless something is done to turn the unit off very rapidly.

This type of unit works very well when the output is rectified to DC again and it will work reasonably well as an AC supply for resistive loads. However, we see that the output voltage is a long way from a sine wave and that the frequency of the output swings widely with a varying load. Commercial units of this type will usually not operate an induction motor because the frequency is too high. Also, the high harmonic content of the wave will overheat induction motors and transformers. Large capacitive or inductive loads will usually cause the inverter to stall or overheat.

You will note that the transistors are operated in the switching mode; that is, they are either cutoff or saturated. A word on that subject is in order. When a transistor amplifier is operated in the linear class B mode, it can deliver very low distortion sine waves to the load. However, in this case the transistors can be shown to be no more than 50 percent efficient. The reason for this is the fact that the transistors have a substantial voltage drop during the rising and falling portions of the sine wave at the same time that they are drawing a substantial current.

The disadvantages of this are not only the fact that one is wasting half of the hard-won electricity but the fact that this electricity is being wasted in the form of heat in the transistors themselves. This must be removed if the transistors are not to be destroyed by overheating. The net result of this is the fact that the transistors can handle a far larger load in the switching mode than they can in the linear mode of operation. For example, let us consider a power tab transistor, the GE D44H11. This unit currently sells for something like $1.50 in lots of a dozen or so and is representative of moderately sized, fast switching, power transistors. The unit has the following significant specifications:

$$V_{ceo} = 80 \text{ Volts} \qquad V_{ebo} = 5 \text{ Volts}$$
Power Dissipation $= 32$ Watts (case heat sunk to 70°C)

$V_{ces} = 1$ Volt (@ $I_c = 8$ Aand $I_b = 0.4A$)

In a push-pull configuration, similar to Fig. 5-1 (except that the base drive would come from a sine wave source rather than the feedback winding), we see that under the stated conditions a class B amplifier could deliver a continuous output of 2×32 W = 64 watts. This is a maximum rating.

In the switching mode delivering square waves, we find that the same transistor could handle the switching of a 24-volt supply (remember, the transformer doubles the supply voltage at the collector) and when turned on, the ON transistor could pass 8 amperes. This yields a power of $8 \times 24 = 192$ watts. In this mode of operation the OFF transistor dissipates nothing and the ON transistor dissipates $1V \times 8$ A = 4 watts. The division by 2 results from the fact that each transistor is ON only one-half of the time. There is also a certain amount of power dissipated in the turn-on and turn-off period since the voltage and current do not fall to zero in zero time. However, with a 1-MHz transistor at 60-Hz this extra dissipation is very small. It does not become significant until the square wave frequency gets up into the kHz region.

From the example we see that the operation of the transistors in the switching mode increases the power handling capability of a given pair of transistors by a factor of three or better and considerably reduces the heating at the same time. Unfortunately, as noted earlier, the square wave output is not very suitable for operating transformers and induction motors. This is true even if it is of the correct 60-Hz frequency because of the heating problems.

If purchasing a commercially made inverter for home electricity examine the specifications to make sure that the unit output is regulated at 60 Hz and that the output is reasonably close to a sine wave. Harmonic content up to five percent is usually acceptable. Also, check the range of power factors over which operation is guaranteed. Without these precautions you may purchase a unit which is not useable for all of your requirements.

From this we see one of the basic difficulties of building a solid-state inverter. The common variety of inverter does not easily supply sine waves and is very sensitive to load conditions. Both of these problems are easily handled by the motor generator set which inherently tends to furnish reasonable sine waves and has the energy of the inertia stored in the rotors (flywheel action) to supply a portion of the momentary overloads.

It seems logical to ask whether there is not some form of inverter which can take advantage of switching-mode operation for efficiency and cool operation of the transistors while at the same time providing a more reasonable approximation to regulated 60-Hz sine waves. We shall consider a few of the possibilities.

THE QUASI-SINE WAVE INVERTER

One of the first possibilities to suggest itself is that the sine wave might be approximated by a suitable staircase voltage. Obviously, the sine wave can be approximated to any required degree of precision by a number of steps if the number of steps is allowed to be sufficiently great. However, from an economic viewpoint the question becomes one of deciding the number of steps which requires the smallest number of transistor switches. And yet, these steps must still provide an approximation of the sine wave which is low enough in harmonic content to avoid some of the heating effects previously discussed.

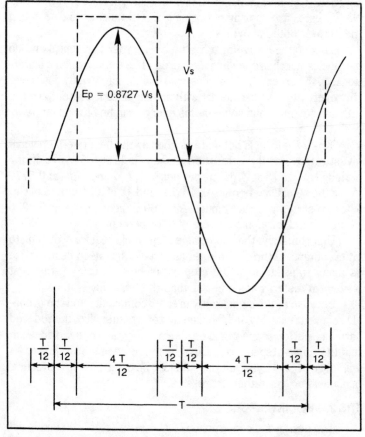

Fig. 5-2. The one step quasi-sine wave.

Rather surprisingly, the first such happy approximation comes with a single-step, judiciously timed. To the best of the writer's knowledge, an algorithm does not exist to permit the mathematical solution of an optimum waveform with an arbitrary number of steps. However, the harmonic content of any particular waveform is solvable by means of the Fourier Transform which can be implemented on a computer. The waveform on Fig. 5-2 was selected from a computer search. If the waveform shown dotted is compared to the largest sine wave component it may be seen that the V_s is just 14 percent higher than the largest sine wave component. If the wave period is considered to be divided into 12 equal parts, this waveshape could be obtained by turning Q_1 ON during periods 2, 3, 4, and 5 and turning Q_2 ON during periods 8, 9, 10, and 11.

A circuit for providing this waveshape is shown in Fig. 5-3. A 7492 divide-by-twelve counter is used to drive a 74154 4 to 16 line decoder. This unit is an active low device. Therefore, the 4 wide NAND gates are actually performing an OR function and selecting the periods noted above.

This voltage waveform has no even order harmonics and no harmonics which are multiples of three. The total harmonic distortion is 7.45 percent. This is somewhat higher than strictly desirable. However, the first harmonics with any significant content are the fifth and seventh, and some of this energy can be filtered out fairly easily.

This waveform is not quite as good as the best two-step wave investigated. For the latter the voltage is held at 0.5774 V_s during periods 1, 2, and 3 at V_s during periods 4, 5, 6, 7 and 8 and at 0.5774 V_s for the remaining periods 9, 10, 11, and 12 of a 24-period cycle. The two-step wave has a fundamental (60 Hz) component of 0.8830 V_s and a total harmonic content of 4.74 percent.

Unfortunately, the odd voltage step would require switching to different taps on the transformer and would thus need nearly twice as many transistors. If the step is selected as 0.5 V_s, the best waveform tested was no better than the best single-step.

Because of the significantly greater complexity and the probable reduced reliability and the question of whether all of the reduced harmonic performance would be obtained in practice, it would seem that the single-step inverter might be the more practical choice. As will be seen shortly, even the simplest inverter built in a practical household size is rather impressive.

THE 2.6 kVA INVERTER

The design and construction of a large inverter is definitely *not* a good project to be undertaken by a rank amateur with little in the

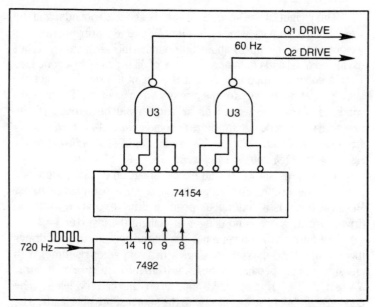

Fig. 5-3. The single step quasi-sine wave drive.

way of test equipment. If you are not reasonably well equipped and knowledgeable in electronics, the best course is to shop around and *buy* an inverter. The following design is presented mainly to give a feel for what to expect and to serve as a guide for the more experienced.

The circuit diagram of Fig. 5-4 shows the inverter circuit. In order to pass the required current it is necessary to parallel 13 of the D44H11 transistors. In the collector of each is a low-resistance resistor to prevent current hogging and to dampen spiking a bit. It is important that this resistor be non-inductive. Base drive for the parallel groups comes from a PNP transistor. The base drives are equalized by the 28-ohm resistors. The PNP is in turn driven by the Darlington pair which brings the input sensitivity down to TTL level.

Transformer T₁ is an Arnold AA533 Silectron "C" core weighing 18 pounds. The secondary consists of 131 turns of B&S No. 2 gauge copper. The two primaries consist of 39 turns of B&S No. 0 gauge copper with a tap at the 33T point for operation at low voltage. The core is made of 12-mil laminations in deference to the harmonic content. If the transformer is immersed in a good transformer oil, the inverter is capable of continuous duty at 2.6 kVA. Note: When packing a transformer in oil it is a good idea to pull a vacuum of about 27.5 or more inches mercury to assure that all bubbles are withdrawn from the windings.

The weight of the secondary wire is about 20 pounds, and the weight of the primary winding is about 22 pounds, bringing the total weight of the unit up to about 60 pounds, including the core but exclusive of oil and container. The use of oil is probably a good idea from a cooling standpoint since it is a bit of a struggle to get the appropriately sized wire for the primary on an inverter. The primary windings should be wound bifilar and should be outside of the secondary in this design for better cooling. *Note: An inverter of this size produces lethal voltages. The unit should be handled with the respect given high-voltage equipment.*

The unit as shown should handle about 26 A on a continuous basis on the output and surge to 33 A on a one-minute basis. Because there is a continuous positive drive excitation from the driver circuit, there is no danger of stalling the inverter.

The filter on the power input is required because the unit draws its current in 5.56-millisecond slugs with a very steep risetime. With any significant impedance in the leads from the battery or generator, a great deal of ripple would appear on the battery leads. The condensers should be installed with the shortest possible leads. The choke should have sufficient airgap to permit non-saturating operation up to 150 A.

Care should be taken in heat sinking the transistors. Each of the large units will be dissipating about four watts maximum. The unit should be safely operable up to a heat sink temperature of 70+C.

The A-B switch in the primary is there to permit operation from either the batteries alone or from the batteries with the generator; i.e. 12 volts or 14.2 volts. This switch should only be operated with the unit non-operational.

The pair of diodes CR_1 and CR_2 are present for the free-wheeling function. They are necessary since the instantaneous interruption of the primary current could give very large spikes which would destroy the transistor bank if the unit is operating without secondary load. These units should be rated in excess of 150 amperes and should have the highest switching speed possible.

RADIO FREQUENCY INTERFERENCE

The entire unit should be completely enclosed in a metal case with all leads coming through RFI suppressors. The construction of the case should be either all-welded (except for the entry door) or riveted or bolted with a fastener every lineal inch along the seams. Without this precaution, neither you nor your neighbors will be able to listen to radio any time the unit is in operation, and the TV and FM will also be troubled by the RFI.

LARGER TRANSISTORS

Several firms currently offer transistors with higher ratings than those selected. For example, General Semiconductor Industries is offering a T0-3 packaged unit rated at 50 A and 200V. Starting with a 56.8V input, these units could be used singly in the opposite sides of the inverter. Unfortunately, they are considerably more expensive, and inverters have a way of chewing up transistors during the development and test phases.

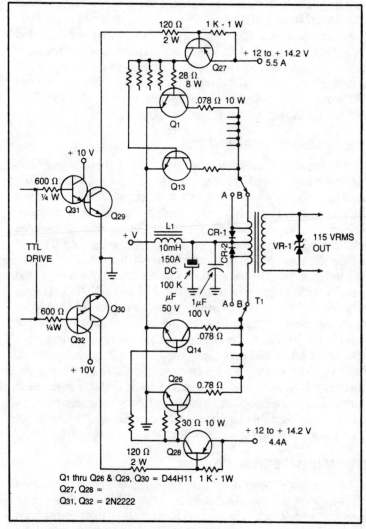

Fig. 5-4. The 2.6 kVA inverter.

As noted earlier, the design and development of a large inverter is not a task for beginners. The internal layout of the unit should permit the use of the shortest possible leads, and the leads from any of the parallel units should be the same. Depending upon the mechanical design, the total weight of the unit with case and transformer and filters will run around 250 pounds.

A particularly tender point in a design of this type is the loss of the output load at the instant when one of the current pulses ceases. If the load has been heavy and the current in the inductor primary is built up to a high level, a large transient can result on both the input and output. Item VR_1 is a GE MOV varactor intended to limit this transient. These units function very much as if they were a pair of back-to-back zener diodes. When the threshold is exceeded, they draw current heavily and tend to absorb the transient energy. For a unit of this size, the spike could contain an energy on the order of 180 joules. A recommended unit would be the GE type V13-0PAECG 20A with six parallel units required.

THE TRANSFORMERLESS INVERTER

As higher voltage, higher current transistors become available, the possibility of implementing the waveform for the single-step approximation without the use of a transformer becomes increasingly attractive. Figure 5-5 depicts this arrangement schematically. The assemblies at Q_1 and Q_2 represent photon-isolated switches which simply switch one end of the battery string onto the load during the appropriate ON period. Complete packaged photon-isolated switches are being offered by a number of vendors, and many of these can be driven directly at TTL level. Note that in this arrangement the individual switch must be capable of holding off twice the supply voltage plus any inductive turnoff transient not absorbed by VR_1. At the time of this writing, switches adequate for a substantial household rating are not available in packaged form. However, the ratings are creeping up on these levels and the possibility exists that such units may become available in the near future. In all likelihood, this simple switching arrangement with commercially available packages will be far cheaper to construct than a more common inverter or a motor generator set.

TRUE SINE WAVE INVERTERS

A number of vendors have developed some very sophisticated true sine wave inverters for "un-interruptable" power supplies to be used with computers. The requirement for these units stems from the fact that many computers will lose the complete contents of the

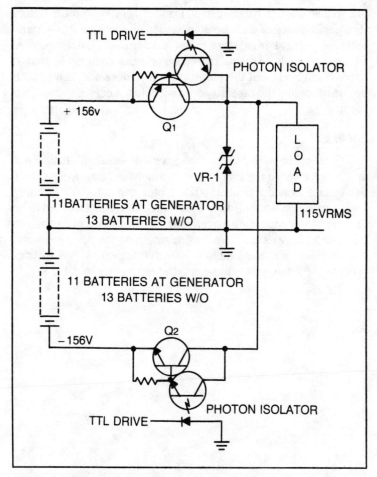

Fig. 5-5. The transformerless inverter.

read-write memory with even a brief interruption of the line power. These special power supplies are built to fall back upon battery storage in the event of failure of the line voltage. Many will do this without even so much as a phase glitch on the line into the computer.

Most of these units operate at a frequency which is a relatively large multiple of the line frequency, such as 20 or 30 times. The sine wave variation is generally obtained by sine wave modulation of the duty factor either by pulse width modulation or by pulse repetition rate modulation. The high frequency is rectified and filtered to obtain the low distortion of 60 Hz sine wave.

These units tend to be very expensive, and a discussion of the design is well beyond the scope of this text. In general these units are well adapted to handling fixed loads of very high power factor. The performance of such a unit in handling heavily inductive transient loads, such as an induction motor starting cycle, is questionable. There is probably no special reason why the unit could not be built to handle such loads, but the prime application does not require such performance and the capability has probably not been designed into the unit.

SUMMARY

It seems to be that the motor-generator set is still attractive in applications where the load may vary widely and may present poor power factors and transients. The motor generator is inherently equipped to produce reasonable sine waves and is not inclined to be "tender" about momentary overloads. As the ratings of transistors increase and prices fall, inverters approaching the robustness and output quality of motor generator sets may appear. At this writing, though, the edge seems to belong to the motor-generator set.

6

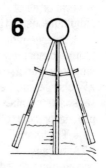

System Control and
Power Factor Correction

From the discussion in the preceding chapters it should be apparent that some very substantial economies are available without sacrificing utility if the home electrical system is rigidly controlled to prevent the occurrence of very high peak-to-average ratios and the existence of bad power factors. In this chapter, we shall discuss some of the mechanisms available to permit control of the system in such manner that these effects are limited.

Obviously, a number of simple common-sense moves can be made without any requirement for additional equipment. For example, the clothes may be dried on the line out-of-doors without using *any* electricity, rather than using the electric clothes dryer. Of course, a wind, rain, snow, or sleet storm may make this impractical, and the indoor clothesline may have to be pressed into use. Another matter is as simple as making sure that all unused lamps are turned off. Dishwashers and clothes washers should be run only when there's a full load. Use the small toaster-oven when the capacity of the kitchen range is not required. All of these simple conservation measures can make it possible to save a significant fraction of the electrical energy consumption.

But there is a large class of services around the home which is not easily done without. For example, when the sump is filled with water you would like to have it pumped out. When the thermostat falls below the pre-set temperature you want the furnace to start. When the refrigerator rises above the pre-set temperature you want the compressor to start, etc. And, if you have your own well, you of course want the water pump to work.

You will note that each of the preceding examples is a machine with some level of automatic control. *You* do not normally turn it on and off; it makes that decision for itself on the basis of some internal criterion. In the home electrical system, there is little to prevent several or all of these machines from reaching the decision to start simultaneously, thus creating a very large peak demand.

In large industrial situations, this problem exists also, and the utility companies actually bill not only for the average power consumption but also for the peak kW demand. In some cases, the peak kW charge can be as large as one-half of the monthly bill. For this reason a number of industrial facilities have instituted a microprocessor control system to reduce the peak demand.

Traditionally, the utilities have not billed homeowners for peak kVA because of the expense of monitoring instruments and the accounting procedures involved. Instead they have simply adjusted the rates to homeowners to account for the average cost. As time continues, this may change. However, at present, there is no economic incentive for the homeowner to install expensive peak-load control and power-factor control apparatus as long as he is buying electricity from the utility.

In the situation where the homeowner is making his own electricity, the emphasis changes dramatically. As was noted earlier with the drill press example, a proper power factor correction would have held the peak demand below 5 amperes instead of the nearly 20 required. Also noteworthy is the fact that we might intervene somewhat in the cycling of the automatic machinery so that no more than one big machine can start at any one time.

By these techniques, it seems likely that the homeowner can reduce his peak kVA demand from something on the order of 15 kVA to something more like 3.5 to 4 kVA. Because the difference in price of storage-generation facilities tends to proceed somewhat faster than linearly, this would represent a reduction in initial cost by not less than a factor of 4 and perhaps by a factor as high as 12 or 16. Thus from an economic viewpoint as well as a conservation viewpoint, it makes sense to consider a fairly large investment in control and power factor correcting equipment. Because of the flexibility, the ease of usage, and the rapidly falling costs, we shall consider that one of the various single-board microprocessors will be used as the heart of the unit.

The detailed discussion of the microprocessor and its operation are well beyond the scope of this book. We shall present the various control algorithms in flow chart form only. It will be presumed that the interface between the processor and the system will be through some device such as the Motorola 6820 PIA (Peripheral Interface Adapter) chip. The detailed workings of the processor boards vary so much from one unit to a different type unit that we could easily fill the whole text with nothing but software for different processor chips. It will therefore be assumed that the reader knows how or will learn to program the unit he has.

REQUEST/PERMISSION CONTROL

Obviously, if we wish to ensure that no two large units will start simultaneously, a request/permission system would be the easiest to implement since it can be done entirely external to the appliance. Suppose, for example, that each of the appliances were equipped with a voice and ears. If the temperature of the refrigerator rose above the setpoint, it could ask the controller, "May refrigerator have permission to start?" The controller could then survey the current power consumption and any other requests and reply "permission granted" or "hold." The latter response would indicate that some other machine was in the starting process. The "hold" would be in effect only during the several seconds that the other machine actually needed to start.

In this connection, remember that a number of machines have some continuing requirement for a small amount of electricity in addition to the main load. For example, the electrical timer and the panel lamps on an automatic washer draw a few watts even when the machine is in the "soak" part of the cycle.

A second thing to remember is that different machines will have different levels of urgency attached to a request to start. For example, it may scarcely matter to the refrigerator or the air-conditioner if they have to wait five seconds for another motor to start. Conversely, the furnace bonnet temperature control requesting a start from the blower might easily overheat in a short additional period and abort the heat cycle. Thus, priorities might have to be assigned to the system. For example, the controller might have to know that the furnace heat cycle has started and "hold" all other requests until the blower was safely started.

With these restrictions in mind, let us examine a very simple form of remote Request/Permission control. The circuit of Fig. 6-1(A) shows a schematic version of the circuitry involved. The 2200-ohm resistor in series with the line permits enough current to flow to permit operation of a small timer and such with nearly full line voltage. During this period, the rectifier following the 6V transformer rectifies a small signal of about 5 volts which is sensed by the computer. When the machine attempts to start, the impedance of the appliance falls sharply and the rectified voltage on the ($\overline{\text{RTS}}$) line falls to nearly zero. This is sensed by the computer as a TTL low. If the conditions permit a start at this time, the computer takes the (PTS) line high which closes the relay, shorting the 2200-ohm resistor, and the machine will start normally.

The selection of the 2200-ohm resistor is nominal. It serves only to limit the current and to prevent a start without the permissive (PTS) signal by limiting the current through the appliance and the (RTS) sensing transformer to 0.052 A. Since the transformer is lightly loaded, it will draw only a few milliamperes and the remainder is available to operate the clock or timer or such. Note that in the case of an automatic washer, the voltage would drop far enough to stop the timer, thus "freezing" the washer in the portion of the cycle at the beginning of the "start" command from the internal logic.

In the case of automatic washers for clothes and dishes, solenoids are usually used to operate water valves. These could not draw sufficient current through the 2200-ohm resistor to operate. Depending upon the design of the particular machine, it may be necessary to bypass the (PTS) relay to permit these to operate. This will require some internal modification of the machine. Your friendly neighborhood repairman may or may not understand how to do this and whether it is required.

MICROPROCESSOR-CONTROLLED MACHINES

A special note of caution is in order with regard to the increasingly larger number of machines which use simple microprocessors for control rather than gear-cam-motor timers. In general, the microprocessor requires a continuous flow of current in order to retain its memory. If the electricity is shut off—even very briefly—the processor memory will be wiped clean and the unit will go back to square 1 in the process. For such machines, it is imperative that an internal alteration be made to continuously supply current to the control logic to permit the machine to complete the cycle. Without this provision the machine will get only as far as the first high-current demand cycle and reset to the start condition.

SOLENOID VALVES

Another note of caution is in order regarding solenoid valves. On a number of washing-type machines, a solenoid valve is timed to turn the water ON. The turnoff command is obtained by sensing the point, after the starting surge, when the motor current either shifts phase or increases due to the load of churning the water. If the motor is not allowed to start by the load control, the OFF WATER command will not be issued and the machine will overfill or overflow. Machines using this logic are generally older type designs with mechanical clock timers. Interrupting ALL of the current will usually prevent the solenoid from turning ON; thus, preventing the flood. *On all timed-cycle machines it is advisable to apply the $(\overline{RTS})/(PTS)$ type control only after a careful consideration of the effects of the interruption of supply current upon the cycle.*

SOLID-STATE SWITCHING AND SENSING

The circuit of Fig. 6-1(A) is shown using a relay and a transformer for sensing. It is entirely possible to accomplish both of these functions entirely with solid-state components. The circuit of Fig. 6-1(B) illustrates a solid-state AC switch with a photon isolation feature. The component designated U_{1A} and U_{1B} represents a packaged photo-SCR isolator such as the GE H211C. This unit is capable of operating directly from 115V AC lines and provides a high voltage isolation to the TTL control line. The unit comes in a 14-pin dip package and can handle 0.300 A. This is too small for most appliances large enough to warrant $(\overline{RTS})/(PTS)$ control but is sufficient to permit control of SCRs or TRIACS large enough to do so. The latter should be selected to handle the appliance surge. For example, the drill press motor which draws 19 amperes on start could be handled by a pair of GE C36B SCRs.

The solid-state sensor shown in Fig. 6-1C employs a transistor photo isolator such as a GE H11A-1. The diode CR_2 is any modest size diode, such as an 1N4000. It serves only to protect the light emitting diode in the photo-isolator. These units have a very low peak reverse voltage, typically less than 5V. It should be noted that the photo transistor isolator may not always give a good TTL level ON and it may be necessary to follow the stage with a transistor.

NEGATIVE TRUE LOGIC

It will be noted that the $(\overline{RTS})$ signal is a negative-true logic signal and that (PTS) is a positive true logic signal. This is actually the advantageous way to operate the system since any of the $(PTS)/(\overline{RTS})$ devices may be unplugged from the central controller without effect. In the absence of a low on the line the controller will make no distinction between an unplugged line and a device not wanting to start and will therefore take no action.

In a similar manner, the active HIGH on (PTS) protects the unit controlled. If the unit is unplugged it makes no distinction between the unplugged condition and a HOLD, and will not start. This is important since it is possible to disconnect any unit without having anything start up unexpectedly.

LOAD SHEDDING.

Most of the industrial-type load controllers function to limit the peak demand for electricity to minimize the demand charge from the utility. They generally tend to do this by classifying the various loads into SHEDDABLE and NON-SHEDDABLE loads. They may assign a priority to these loads. Prime examples of sheddable loads are:

electric hot water heaters
sidewalk defrosters
pool heaters
air conditioners.

Such devices can be turned off without any serious consequences except as noted in the following paragraphs.

There is usually a series of non-sheddable loads which should be permitted to operate if at all possible. This includes items such as:

electric lights
computers
cash registers
process controls.

Obviously, it would not be desirable to have the lights go out or to have the computer crash every time the general load got high.

These categories really apply largely to industrial and commercial applications. For the home electric system, the likelihood that one would use his hard-won electricity for items as frivolous as sidewalk defrosters and pool heaters is very small. However, there are always some sheddable loads and also some non-sheddable loads. In addition to these, some deferrable loads and non-deferrable loads exist. Also, some of the loads may be deferrable *and* "non-interruptible."

LOAD DEFERMENT

There is a class of loads which can be readily deferred but may not be interrupted, as for example, the typical forced-air gas furnace. The machine draws a small but steady load from the line to operate the thermostat system. When the room thermostat senses that the room temperature is below the setpoint, the solenoid valve opens and the burner lights. The burner is usually equipped with a non-electrical interlock which prevents the operation of the solenoid valve if the the pilot is not lit. Alternatively, on some units electrical

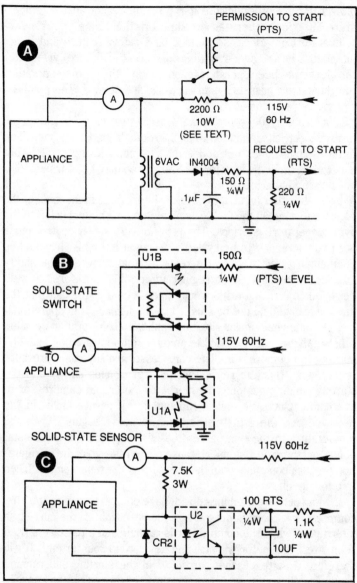

Fig. 6-1. REQUEST/PERMISSION TO START circuits.

ignition is used, and the spark is ignited to light the flame when the solenoid supplies gas. On some oil units, a pump and blower are also started to furnish oil and to blow the flame.

After the flame has been burning for some time, the bonnet or heat-exchanger thermostat gives the motor start command to the blower. If the motor does not start for some reason an overtemperature interlock operates to shut off the flame. This overtemperature interlock is intended as a safety feature and is not intended to be a regular operating portion of the furnace cycle. After the blower has been operating for some time, the room temperature thermostat will pass its hysterisis range and open. At this point the heating cycle is ended.

It probably does not make a great deal of difference if the furnace electricity is shut off while the room thermostat is open. The furnace simply does not sense the closing, and the cycle is not initiated. If the start is deferred, the rooms may get a trifle cooler but nothing is harmed.

On the other hand, if the cycle is interrupted in the middle, the onus for turning off the fuel to avoid melting the bonnet is left to the overtemperature interlock. This is generally a safety control and is not really intended to be cycled at frequent intervals. It would be poor engineering practice to create a situation where the safety interlock were cycled at regular intervals. Therefore, a well-designed control algorithm would take account of the fact that the furnace load should not be shed while the furnace is in operation.

A somewhat similar situation applies to refrigeration devices. Most of these units respond to an over-temperature command by starting the compressor. The compressor usually has a pressure reservoir so that the refrigerant pressure builds slowly and the starting winding has only the inertia of motor and compressor to overcome. Once the unit is running, the pressure builds in the reservoir and the actual refrigeration process begins. When the temperature in the box passes the low setpoint, the thermostat opens and the motor stops. Because of the reservoir, the refrigeration process continues until the pressure in the reservoir has fallen to a low level.

Most of these machines do not have enough torque to start the compressor if there is pressure in the reservoir. If the current is interrupted while the compressor is running and a re-start is tried soon after, the motor will stall and develop no counter emf. It will therefore draw excessive current, and the starting and running windings will overheat. A temperature sensing cutout will usually interrupt the current to prevent the burnout of the motor. When things cool down the cutout will close again and the systems will try another start. Interruption of the refrigerator may not actually destroy anything; however, it is poor engineering practice. The cutout was not designed for regular cycling, and the overheating will significantly shorten the life of the motor. Most refrigeration machines should not be interrupted while running, and if it is necessary to do so, they should be locked out and prevented from attempting a re-start for a sufficient period to permit the reservoir pressure to bleed down to neutral.

SELF-DEFROSTING MACHINES

In passing it should be noted that self-defrosting refrigerators and freezers are notorious wasters of electricity. These machines usually apply an electric heater to cause the condensate to remain liquid and run off and then apply an evaporation fan. This obviously wastes a great deal of electricity. In a few of the machines the self-defrost feature can be disabled. For most home electrifiers, the logical choice will be to defrost the unit manually from time to time and thus save the cost and effort involved in providing the system capability and storage required for this rather wasteful practice.

The burn-out type of self-cleaning oven represents an even more atrocious use of home electricity. Just do not turn on this feature unless you really intend to test the maximum load capability of your system.

LOAD CURRENT SENSING

While it may not be immediately apparent, an important part of the type of load management described is the ability to sense when the deferrable but non-interruptible loads are actually drawing current. The basic principle of the $(\overline{RTS})$/(PTS) scheme requires the cancellation of (PTS) as soon as the device is satisfied. While this can be done in some instances on an elapsed-time after (PTS), in a great many it is either not convenient or not possible to do so.

For example, in the case of the furnace, the length of time that the unit will run can be quite long in cold, windy weather. For the refrigeration devices, the time can be long in hot weather or after the door on a freezer has been held open while stocking the shelves with warm food. The overall system speed could be severely penalized by operating on a worst-case time delay. Conversely, the system can operate with maximum dispatch if the computer can sense the cessation of normal load current. This cessation will tell the computer that the furnace is up to temperature or the refrigerator is down to temperature or the sump is empty. Once the device has shut off, the computer can safely cancel (PTS) without interrupting the load cycle.

The illustrations of Fig. 6-2 show two forms of current sensors. These circuits are installed in series with the load and are represented by the ammeter symbol Ⓐ in the three illustrations of Fig.

6-1. The first of these is an opto-isolator circuit similar to the (RTS) sensor of Figure 6-1(C). When the load is drawing rated current, a three-volt drop is noted across R_s, and the light emitting diode of the photo isolator will be illuminated and the photo transistor switched ON. Diode CR_1 protects the LED from reverse polarity. The small filter is added to ensure that the pulsations from the half-line cycle are not seen by the computer.

The rather large drop and consequent heating in R_s are necessitated by the fact that it requires about 1.75V DC to turn the photo diode ON. This feature makes this form of load current sensing less attractive than the circuit shown in Fig. B.

The circuit shown in Fig. 6-1(B) employs a current transformer similar to the one described in Chapter 1. Since the voltage developed across R_L is proportional to the line current, a low current will not develop sufficient voltage in the filter to bias the transistor ON.

Both of these circuits will withstand a five times normal starting current for a number of seconds. Both also provide a negative logic starting and running signal since unplugging of the unit will be interpreted correctly by the computer as a no-current condition, whereby the computer will cancel (PTS).

THE CONTROL CODE

In order for the computer to be able to distinguish the various forms of treatment to be accorded each individual load, a control code or control word is established. This control code should contain the significant information. The code could be almost any length. However, an eight-bit code is convenient for most of the micro-processors used in a control application of this type. In the example that follows, this format is used.

First of all, let us assign the three lowest order bits, b0, b1, and b2, to the starting time code. These three bits can give us any binary value from zero to seven. If a unit time of 4 seconds is assigned, we can have any starting delay from 0 to 28 seconds. This is adequate for most appliances under most loads.

The next three bits, b3, b4, and b5, can be assigned according to the current draw of the device. For example:

b5	b4	b3	interpretation
0	0	0	less than 5 amperes
0	0	1	5 amperes
0	1	0	10 amperes
0	1	1	15 amperes

	b5	b4	b3	interpretation
	1	0	0	20 amperes
	1	0	1	25 amperes
	1	1	0	30 amperes
	1	1	1	35 or more amperes

Next, we may assign the two most significant bits, b6 and b7, to the device classification. This is as follows:

b7	b6	Interpretation.
0	0	Never interrupt (reserved for lights, and other critical services).
0	1	Defer but do not interrupt (furnaces, refrigerators, etc.).

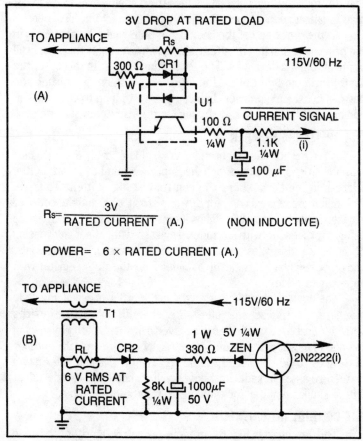

$$R_s = \frac{3V}{\text{RATED CURRENT}} \text{ (A.)} \qquad \text{(NON INDUCTIVE)}$$

$$\text{POWER} = 6 \times \text{RATED CURRENT (A.)}$$

Fig. 6-2. Solid-state current sensor at A and transformer type current sensor at B.

b7	b6	Interpretation
1	0	Defer or interrupt but delay re-start (air conditioners, compressors, etc.).
1	1	Interrupt first (hot water heaters, ovens or other always-sheddable loads).

The purpose of this coding is to provide a scheme with a mathematically determinable value in proportion to the desirability of shedding the load. For example, a device with the code 11111 in the most significant bits (equivalent to the decimal value 31) would be a desirable one to drop in a heavy overload since it is classed as always sheddable and it draws in excess of 30 amperes.

A further value is the fact that any device classed as always sheddable will have a larger numerical value than the largest device in the defer or interrupt but delay re-start category. It can be seen that the code thus organized provides the computer with a clear-cut mathematical route to select which devices are to be dropped.

At the other end of the line, at the time when the system is in the process of starting up after a shutdown, it is likely that nearly all of the automatic devices will be requesting a start. In this case, the system will pick up the load in an orderly fashion by starting the devices in ascending order, building the system up to capacity and adding items after some of the higher priority items shut off.

THE APPLIANCE NUMBER

A further item is required for control and this is the appliance number (APN). In actual fact the appliance number is simply the address in memory where the control code is written. Thus we would find the control code for device zero at x000 and the code for device number 3 at x003. This scheme is particularly handy when using a processor with an index register with offset addressing because the addresses for the code, the input and output terminals, and other requirements can be obtained from the index register with appropriate offsets.

The devices should be arranged in the order of ascending code numbers, but this scheme offers the added flexibility that devices with identical code numbers can be given a further priority by placing the one to be shed first at the higher (APN). In general, the number of devices to be controlled will be relatively small and one or two sixteen-register hexadecimal pages will suffice for the code table, with a similar number for the input/output address table.

THE CONTROL ALGORITHM

The flow chart in Fig. 6-3 presents a control algorithm arranged to provide for orderly assumption of load and orderly restoration after a load shed has occurred. It is assumed that none of the devices coded 00xxxxxx are equipped with either switches or current sensors and that these devices will be shed only manually, as by turning off the lights and so on. The various portions of the program have been labeled by enclosing the title in an ellipse.

Any complex algorithm requires some study in order to obtain a reasonable understanding of the operation and this is no exception. However, to try to make this one a bit easier, let us "walk through" this program on a "guided tour" and comment on the sights along the way. We will begin by assuming a "cold" start when the system has been down and nothing is running.

PROGRAM START is the reset address of the processor. At this address in the ROM (Read-Only-Memory) there are a series of instructions which initialize the processor by clearing all (PTS), flags, and DELAYS. After this, the machine loads the control codes in read/write memory. The loading may be either manual from a keyboard or from tape depending upon the facilities available. The system then sets (APN) to the first 01 coded address. This is the first device to be controlled and enters MONITOR. In MONITOR, the overall system current is first measured and subtracted from the system limit. Any time the measured current exceeds the system limit the program falls into FIND FIRST SHEDDABLE. This program starts at the top of the (APN) list and proceeds downward until the first device drawing current is found. It then obtains the control code and goes into SHED MODE SELECT. On the basis of the control code it will decide what to do about the load. Since we are at startup and no loads have been turned on except the 00 loads, the device would fall through the decision tree and wind up at the extreme right giving the "00" alarm. This would indicate that the non-sheddable loads were exceeding the system limit and that something must be manually disconnected.

In the case where the system load was within tolerance, the processor would obtain the I/O address of the first 01 (deferrable but non-sheddable) device and determine whether it was requesting a start (RTS). It is rather likely that at startup the device would be requesting a start since the system has been down. In this case, the control would proceed to the PERMISSION TO START routine.

In PERMISSION TO START, the processor would first obtain the control code and compare the current rating of the device with the margin between current being drawn and the system limit or capacity. If the system capacity would not be exceeded, the processor would set the start delay taken from the CONTROL CODE and would initiate the start. When the start delay has timed out, the

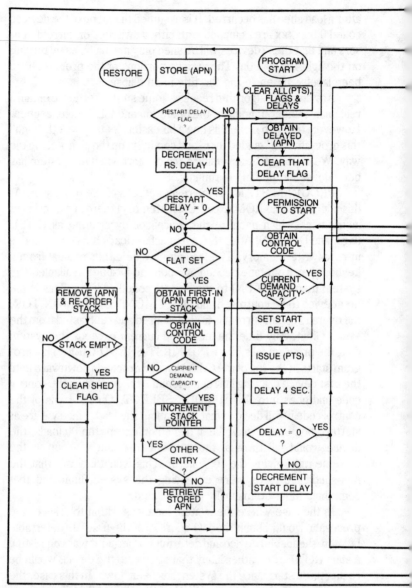

Fig. 6-3. A control algorithm.

control jumps back to the start of MONITOR, re-setting the value of (APN).

On the second time through, the first device will no longer be requesting a start. Therefore, control will fall through (PTS) CAN-

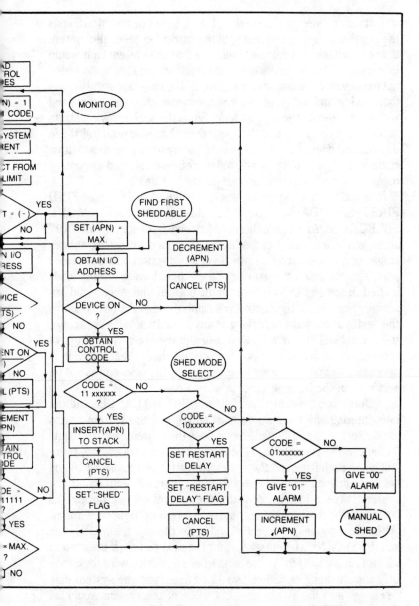

CEL. If the device were not drawing current, (PTS) would be canceled at this point. In any event, (APN) is incremented and the new control code compared with 01111111, which is the code for the last deferrable, "non-interruptible" load, and the next device will be tested for a start code after passing through MONITOR.

It may be seen that all of the 01 devices will be added as fast as the starting delays will permit until the current exceeds the system limit or until the current code indicates that the system limit would be exceeded. In the case where the system current is not exceeded but the device code indicates that it would be, the system will bypass that device and will search for a device which could be started without exceeding the current limit. Since the search always starts at the bottom of the 01 list, the control will always be polling all these devices and will cancel (PTS) as soon as any device ceases to draw current. A high priority device will therefore be started as soon as doing so will not exceed the system current limit.

In actuality, this bypassing would involve a pass through FIND FIRST SHEDDABLE and a fall through the SHED MODE SELECT tree. At startup, this would involve issuance of a brief "01" alarm and an increment of the (APN). This loop will assure at later stages of the operation that whenever a high-priority load is denied permission to start, at least one of the active, lowest-priority loads is shed. Because (APN) is reset to the bottom whenever a load has actually been shed, the control will climb through the (APN) list to the device and try again. As long as any 11 or 10 devices are active, the control will continue to shed load and try again. At startup, of course, there are no 11 or 10 loads to shed. Therefore, it is necessary to bypass the device. At startup, the devices will not be serviced on the basis of strict priority alone.

Only when the control can fall through (PTS) CANCEL without encountering any requests to start ($\overline{\text{RTS}}$) and the system begins encountering 10 and 11 addresses, does the system test for 10 and 11 flags as the result of load shedding. At startup, of course, there will be no such flags and the system will attempt to start all 10 and subsequently all 11 loads requesting a start. Eventually, if the system current limit is adequate, the control will fall through all (APN)s without a start request and the system will be in equilibrium. When this happens, the control will cycle through the MONITOR and (PTS) CANCEL loop waiting for the next ($\overline{\text{RTS}}$). If the system load is running fairly light, a request from a 01 device will necessitate only the shedding of one or two 11 devices, and the system may never get into the 10 devices. In this case, the system would fall through RESTORE to the load SHED FLAG test. The second RESTORE loop is set up to find the load that was shed earliest, which will not exceed the current limit. The arrangement is intended to provide for the fastest possible return of any load which has been shed without regard to priority. If no SHED flag is set, the control polls the 10 and 11 loads in order of priority.

If a 10 device has been shed, the control decrements the delay register and tests to see whether the delay has timed out on the shed 10 devices. When the timeout is complete, the device is cleared for re-start. It should be noted that this timeout is somewhat variable since any devices being started during the poll may have a start delay which will slow the circuit of the loop. Since this makes the delay only slightly longer it does no harm to the system.

It will be noted that the operation of the RESTORE loop is such as to advance the priority of a previously shed 11 device ahead of all but the first 10 device and that would get precedence only if it had been shed earlier. If the system is operating reasonably within its capabilities, the shedding of a 10 load should be relatively infrequent and the servicing of previously shed 11 loads in precedence is probably proper. Too frequent shedding of 10 loads is a sign of system overloading. At this point either the use of some of the appliances should be discontinued or the system generation and storage facilities should be enlarged.

The algorithm shown here may not be optimum for all applications and may need to be enlarged or reduced to fit your requirements. Certain "bells and whistles," (i.e. day/night re-prioritizing and active keyboard service to permit changes in parameters such as the system current limit without interrupting operation) have purposely been eliminated for the sake of simplifying an already complicated situation.

This algorithm should run fairly fast on most control oriented processors. If no $(\overline{RTS})$ is encountered, the complete circuit should run in a few milliseconds. It may actually be necessary to place a few delays in the algorithm to permit the line a chance to stabilize. For control oriented processors, the entire firmware can probably be coded in 1K bytes of ROM and the scratchpad memory requirements should not exceed 128 Bytes if the algorithm is written at the machine language level. For high-level language machines such as the "appliance" computers now on the market, the requirement for memory would be considerably larger. Most of these machines would not have enough I/O lines to handle the various switches and sensors without an additional I/O board. Each 01, 10 and 11 item will require three lines for $(\overline{RTS})$, (PRS) and $(\overline{i})$. A total of 16 controlled items would require 48 I/O lines plus the lines needed for display, if any, of system current sensing and such.

POWER FACTOR CORRECTION

The importance of having the power factor of the system high was discussed earlier in the discussion of the motor-generator and generators in particular. We noted that the lagging current drawn by an inductor or an induction motor could cause the voltage of the alternator to sag, that a leading phase angle from a condenser could cause the voltage to soar, and that a proper power factor would maximize the capability of the system. In particular we showed in Fig. 1-4 that the correct capacitor would reduce the current drawn by the drill press motor from something over five amperes to something under one-half ampere in the no-load condition. For the small home electrical system, a reduction from 585 VA to 56 watts is obviously worthwhile.

One might consider the possibility of correcting each motor individually for something like an average load condition since the power factor tends to change with load. This works reasonably well. However, it does not compensate for the starting surge which is much larger than the running reactive volt-amperes. We shall discuss a system which makes it possible to dynamically compensate the system power factor.

THE PHASE DETECTOR

The circuit of Fig. 6-4 is a sine wave phase detector similar to the circuit discussed with the phase-locked loop in Fig. 4-7. The transformer can be similar to the current transformer described in Chapter 1. It can be seen that the magnitude of the voltage at A is the vector sum of the sample of the line voltage and the current sample which is in quadrature with the line current since the secondary of the transformer is very lightly loaded. The difference between the magnitudes of voltages A and B can be shown to be directly proportional to the amount of the reactive current being drawn by the load. The real parts of the line current cancel and the voltage samples are constant. Therefore, a DC analog-to-digital converter sensing the voltage difference between terminals A and B will provide a digital reading directly proportional to the item which we wish to cancel, namely the reactive amperes. Since we know the line voltage to be constant, our ccntrol computer could easily take this reading and calculate the amount of capacitance required to compensate the power factor.

THE CAPACITOR BANK

The illustration of Fig. 6-5 shows a suitable capacitor bank for power factor compensation in a home electrical system. You will note that the capacitors are arranged in binary sequence of values ranging from 20 μf to 640 μf. By selecting proper combinations in

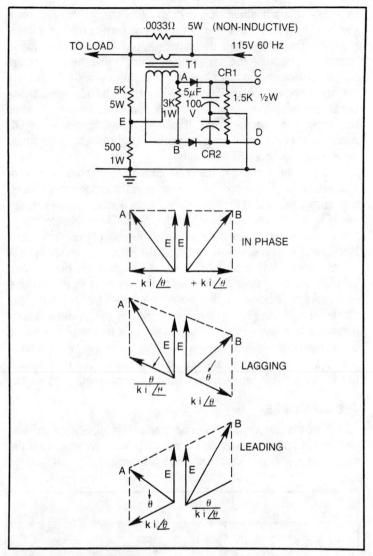

Fig. 6-4. The phase detector.

binary sequence the bank can draw anything from 0 to 55 amperes leading in 0.87 ampere steps.

In the figure, each capacitor is equipped with a double-throw switch going to a discharge resistor. If a relay or other mechanical switch is used to insert the capacitors, there is a fair probability that the capacitor will be disconnected from the line at a time when the capacitor is fully charged. If the switches are actually solid state switches, it is a mathematical certainty that this will happen since the SCR or TRIAC will always turn off at zero current, which is precisely the time when the capacitor is fully charged. This is undesirable because the next switch-on might find the line voltage on the opposite swing and a large current surge would take place in the first half-cycle. Accordingly, a second switch and a bleed resistor are provided to make sure the capacitor discharges rapidly after being removed from the circuit. If solid-state switching is employed, make sure that the switching is synchronized with the line voltage zero crossings to avoid transients.

It may be seen that the A/D converter is itself capable of generating the proper codes for insertion of the capacitors provided that a proper scaling is provided for voltages A and B. However, the use of the computer lends some further flexibility to the situation in that it can be used to anticipate the reactive surge of the motor start. Because of the requirement to obtain the magnitude of voltages A and B, there is a filter network in each. This network must have a delay of several cycles. This causes the correction to lag, and there is an initial large transient for a few cycles at each motor start. If the computer controller knows in advance that it is going to issue a start command to a given motor, it can easily look up the power factor correction required for that start and issue the power factor correction data to the capacitor bank simultaneously with the (PTS) issued to the motor. This can be used to minimize the starting transient.

THE CAPACITORS

Note that the capacitors to be used in this application must be units rated for continuous duty across a power line of at least the voltage in use. High voltages and high temperatures greatly reduce

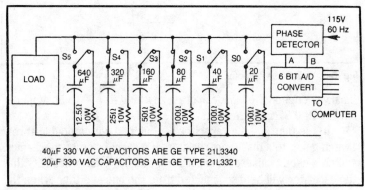

Fig. 6-5. The capacitor bank.

the service life of the capacitors. The units noted on the drawing are rated at 330 volts. They are each a paper and oil type capacitor in a sealed metal case. Current cost for this capacitor bank, if purchased individually, is approximately $500 without the switches or A/D converter.

ECONOMICS

If we assume that the home electrifier has some considerable experience in electronics, the economics of the situation stack up about as follows:

Microprocessor kit	$ 200
Input/Output board	150
$(\overline{RTS})/(PTS)$ & $(\overline{i})$ units	
$45 each times 10 devices	450
A/D converters 6 bit, 2 required	50
Miscellaneous cases, devices, etc.	200
Power Factor capacitors	500
Capacitor switches	150
Total	$1700

This cost assumes that the electrifier will work on the project without charge. At the present reading commercial load controllers start at $3000, and there are no available power factor correctors since there is no market for these devices among people who buy home electricity from the utility company with no penalty for poor power factor.

For the home electrifier the choice lies between relatively sophisticated and expensive load control and power factor correction or the simple brute force approach of making the system large and robust enough to withstand the load transients. Our "typical" average household usage of 1.16 kW can probably be handled by machinery capable of 3 to 4 kW with the sophisticated control and minimum restriction of service or by a 15 to 20 kVA system with little if any control of loading and power factor. Depending upon the nature of the system and the requirement for energy storage, not all of the machinery may have to be enlarged. However, the difference in motor generator size alone is probably a penalty in excess of the $1700 figure.

7  The Windmill

The invention of the windmill is lost in antiquity, and perhaps the windmill is one of those things that was invented a large number of times in a large number of places. Windmills of various sizes have spun in places as diverse as ancient Persia, Tibet, Portugal, and Holland. It seems likely that the Tibetan Prayer Wheel should be accorded the honors for the oldest design. However, a great many ancient types still revolve in various parts of the world. Some of these types are of interest to us here because of the very simplicity. They can be built with the simplest of tools and materials. We shall examine a number of types to see how well they are suited to satisfying the appetite modern man has developed for energy.

After the development of agriculture, there was a desire to grind grain in order to remove the starchy edible portion from the indigestible cellulose husk. This can be done with a mortar and pestle by hand but it is arduous work. As early as the time of Samson (*Judges* 16:21), we find mention of a rotary grist mill with slaves plodding in a circle to rotate the millstones. It is little wonder that men sought cheaper and more efficient means of performing this work. It is also very logical that the machines developed in land-locked areas should be machines of different form than those developed in maritime lands where the use of the sail to extract energy from the wind was well known. The basic difference between these machines is in the direction of the axis of rotation. The landlocked machinery tends mostly to rotate on a vertical axis like a merry-go-round whereas the sail type machines generally rotate on a horizontal axis like an airplane propeller. For the reader interested in the history and lore of the windmill I recommend Volta Torrey's *The Wind Catchers,* published by The Stephen Green Press of Brattleboro, VT.

Substantial differences evolved in the design of the windmill because of the eventual use of the mill. When the object was to grind grain or to pump water from a deep well, it was naturally desirable to have the mill turn relatively slowly to minimize or eliminate the requirement for gearing. The production of a sturdy and long-lived gear from wood is no simple task and requires a great deal of both knowledge and craftsmanship.

The industrial revolution in the United States brought a change in the nature of requirement for windmills. While the steam engine began to take over the requirement for large power in commercial enterprises, the expansion of the American West brought a large market for windmills to pump water and grind grain on isolated farms and homesteads. This in turn launched a very substantial industry which, at its peak in 1910, employed about 20,000 people in the US alone. There were a tremendous number of patents issued for improved wind engines, and a great many home experimenters built their own machines. After the end of World War I, the windmill industry declined. The interest in windmills fell to a low ebb except for a brief spurt in the 1930's when farmers became interested in a new form of electricity-generating machines which would power a radio on a non-electrified farm. These were a different breed of machine which turned much faster because of the high-speed requirements of the electrical generator. As Torrey mentions in his book, the construction of an electric wind engine is as different from a grain-grinding machine as a speedboat is from a barge.

THE WIND

To begin with, let's consider a few of the known properties of the wind to see what effect they will have on the design of a useable windmill. The U.S. Weather Bureau has spent a great deal of effort in recording and analyzing wind patterns and velocities, and these can tell us a good deal about what we can expect to achieve and what we might have to put up with in a wind driven engine.

Experience indicates that there are well-defined groups of winds which predominate and may be called the *prevalent winds*. There is a second group which contain the bulk of the energy which may be called *energy winds*. The reason for the distinction is the fact that the energy present in a column of wind in front of your windmill is proportional to the cube of the velocity. This is a very rapidly rising function as may be seen from the following table. I have normalized the table to 15 mph for reasons to be discussed later.

Table 7-1. Relative energy in relation to wind velocity.

WIND VELOCITY	RELATIVE ENERGY
5 mph	0.037
10	0.296
15	1.0
20	2.37
25	4.63
30	8.0
35	12.7
40	18.96
50	37.0
75	125.0
100	296.3
125 mph	578.7

From the table you can see that an increase in wind velocity from 15 mph to 20 mph more than doubles the available energy, and a decrease to 10 mph nearly quarters it. Weather bureau measurements indicate that energy winds, even in calm summer months, produce about 70 percent of the energy even though they blow only about 42 percent of the time.

Furthermore, the following facts obtain: Energy Winds blow two out of seven days and prevalent winds blow five out of seven days. The mean prevalent wind velocity is 2 mph less than the average monthly velocity. The wind of the highest energy is about 10 mph faster than the most frequent wind. The total energy per month is about twice the energy which would be computed using the average monthly velocity. If the windiest and the calmest months are compared, the energy for the windiest will be about 4.5 times as great and the wind velocity 1.65 times as great. The true velocity will depart from the average velocity by ± 28 percent in a 10-second period and by ± 45 percent in a 38-second period. The high-wind regions of the US (having 10 mph or more average yearly velocity) are a 350-mile strip north and south through the center of the nation, the Great Lakes drainage basin, the Atlantic Seaboard and the Pacific Coast near San Francisco and Washington. The fastest recorded wind in an area is likely to be about 9 to 10 times faster than the yearly average.

Over unobstructed terrain, the wind varies with height with the relationship:

$$\frac{V}{V_o} = \left(\frac{H}{H_o}\right)^n$$

where $n = 1/5$ from 5 to 35 mph.

A telephone call to your local weather bureau will probably yield a yearly average wind velocity at the weather station and factors such as the highest recorded wind. However, the first logical step for anyone seriously contemplating the erection of a wind engine is to put up an anemometer and begin the rather lengthy task of accumulating data for the specific location. As we shall see shortly, the design of the machine you intend to use can be considerably influenced by the wind velocities expected in the area.

WINDMILL REQUIREMENTS

Once written, the following rules are likely to seem self-evident. However, neglecting to observe these rules is responsible for the overwhelming majority of windmill failures, which can result in property damage and injury as well as loss of investment. To be effective and safe a windmill should be:

- Equipped with a brake to permit stopping of the rotor.
- Be designed to operate in the expected energy winds.
- Be equipped to respond to shifts in the wind.
- Be designed to withstand gusts in operation.
- Be designed to survive maximum wind velocity, without blowing down or disintegrating from centrifugal force due to over-speed.

On all but the smallest windmills, the failure to observe these precautions in the design of the windmill will almost certainly result in disaster.

THE VERTICAL AXIS MACHINE

The vertical axis Persian type windmill is thought by many to be one of the oldest designs since it is so elementary. As shown in Fig. 7-1 in top view, the rotor consists of four or eight flat paddles mounted on a vertical shaft. A small building is constructed to shelter a little more than half of the wheel from the prevailing wind. The structure also serves to provide a top bearing support. Obviously, with a machine of this sort, the tip speed cannot exceed the wind velocity. In fact, it will usually run more slowly because of the air being churned by the unused half. It is also fairly obvious that the machine is of use only in areas where the strong winds are fairly fixed in direction. When it was necessary for lubrication or service, the machine could be stopped by closing the opening for the wind.

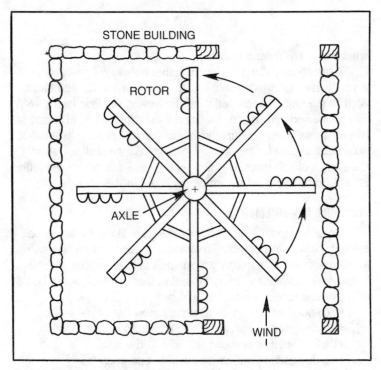

Fig. 7-1. The Persian windmill.

A more modern version of this type machine is found in the anemometer shown in Fig. 7-2. In this machine a set of hemispherical cups is arranged around the periphery of a circle. The open cup facing toward the wind has about four times the drag that the cup facing closed-side toward the wind presents. Therefore, the entire assembly may be exposed to the wind. This arrangement is commonly used for the measurement of wind speed since it does not need to be faced or aimed into the wind. Here again the tip speed is something less than the wind velocity.

The Savonius windmill shown in Fig. 7-3 is a member of this same family. This may be made two- or four-bladed. In the two-bladed format, it is felt that the wind spilling along the flat side increases the efficiency. In the four-bladed configuration, it is more usual to leave the central portion as an open frame. A popular American version of this form was constructed using 55 gallon oil drums cut in half parallel to the axis as the cups. This provided a neat-but-not-gaudy homebuilt design.

For machines with cylindrical cups, the air drag with the open face toward the wind is about 1.3 times the drag of a flat plate of the

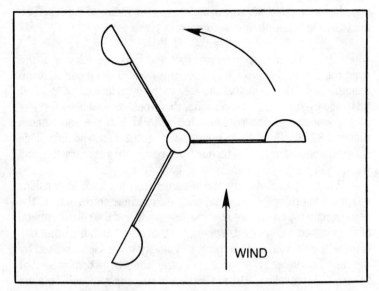

Fig. 7-2. The anemometer.

same area, whereas the drag of the cup with the closed face toward the wind is about a third of the drag of a flat plate. The net static torque is therefore about equal to the torque that would be experi-

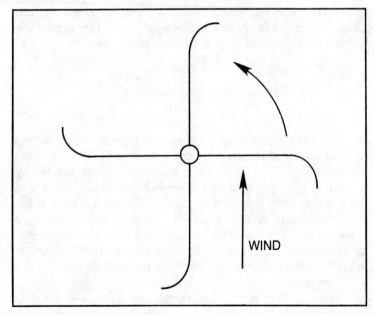

Fig. 7-3. The Savonius windmill.

enced by a flat plate with the same area as the radius of the cup. The pressure on a flat plate due to wind is given as:

$$P = 0.0042 \, V_r^2$$

where P is in pounds per square foot and V_r is the velocity of the wind relative to the surface in miles per hour. Taking our oil drum example, let us suppose that the area of the open face is 2.5×4 feet or 10 square feet. In a 30-mph wind, this would experience a force of 37.8 pounds, with the rotor stalled or locked. If the rotor had a radius of 8 feet, the stalled torque would be 302.4 foot-pounds. This is a tremendous amount of torque, but we are not yet extracting any work from the machine because it is not turning.

When we consider the machine turning, we have to consider several other things. First of all, once the machine starts to turn, the working cup (open mouth) is going downwind and the relative speed of the wind decreases by its speed. On the other hand, the other cup is travelling upwind and the relative velocity it sees is increased by the rotational velocity. If we consider the wheel to be turning so that the cups are travelling 10 mph for the working cup we would obtain:

$$F = 4/3 \times 0.0042 \times (30 - 10)^2 \times 10 = 22.4 \text{ lb}$$

and for the dragging cup we obtain:

$$F = 1/3 \times 0.0042 \times (30 + 10)^2 \times 10 = 22.4 \text{ lb}$$

Since these forces are opposed to one another, it says that this windmill would be yielding zero net torque at 10 mph rotational speed. The more mathematically inclined reader will see that the speed of the mill is firmly fixed by the square root of the ratio of the forward to the reverse drag of the cups or:

$$\frac{V\text{max}}{V \text{ wind}} = \frac{\left(\dfrac{(DF)}{(DR)}\right)^{1/2} - 1}{\left(\dfrac{(DF)}{DR)}\right)^{1/2} + 1}$$

At a cup speed of 5 mph, the working cup will be developing a force of 35 lb and the dragging cup a force of 17.15 lb for a net difference of 17.85 lb. The machine would have a torque (neglecting friction) of 142.8 foot-pounds and would be developing a power of 0.238 horsepower. It would be turning at 8.75 rpm. You can see that this is not exactly the ideal machine for generating electricity when we would like to have our alternator spin at 3600 rpm or more for an automotive alternator/rectifier set. The machine is well suited for heavy pumping and grinding where the tremendous static torque would be of value.

One is inclined to question whether there is not some way in which the tremendous static torque of the machine could be preserved while the D_t/D_r speed limitation is defeated. One such proposition would be the epicyclic machine shown in Fig. 7-4. Each cup is held by a jointed arm and pivots on the end of the arm. The arrangement permits the dragging cup to travel upwind at a very small fraction of the downwind speed of the working cup, thereby minimizing the drag loss and also the drag torque. The balancing and construction of this machine would be no trivial engineering feat, and the machine still could not run as fast as the wind. However, it would be so wondrous to behold in operation that it seems a shame that someone does not take the effort to build one!

There is another class of vertical shaft machines referred to as the Darrieus machine, which resembles a giant eggbeater. These machines can operate faster than the wind and do not require steering. They will be discussed later under the heading of aerodynamic machines. These simpler machines are essentially aerostatic in nature and drift with the wind in much the same way that the square-set sail on an ancient galley did.

AERODYNAMICS

The science of aerodynamics is extensively based upon the principle announced by Daniel Bernouilli in 1745. In essence, Ber-

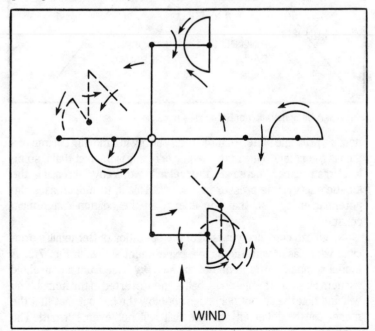

WIND

Fig. 7-4. The wonderful epicyclic vertical shaft windmill.

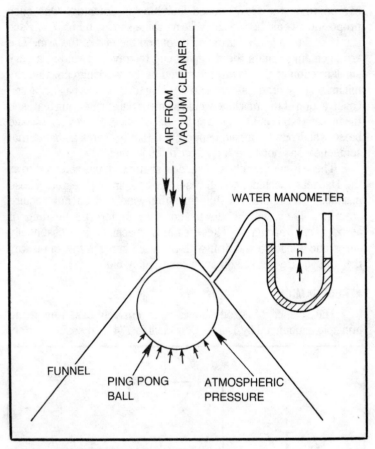

Fig. 7-5. A demonstration of Bernouilli's principle.

nouilli's principle states that the total energy in a moving column of a
fluid is a constant. One of the sequelae of this is the fact that if some
local disturbance increases the velocity, thereby increasing the
kinetic energy, the pressure must fall. This in turn decreases the
potential energy so that the energy in the column can remain
constant.

You can convince yourself of the operation of Bernouilli's principle very easily with the simple experiment shown in Fig. 7-5. A
funnel is taped onto the end of the *blowing* hose from a household
vacuum cleaner. If a pin-poing ball is then inserted in the funnel, you
will find that the air not only does *not* blow the ball out, but that the
funnel can be inverted and the ball will not even fall out! The
explanation is that the air rushing through the gap between the ball
and the funnel attains a lower pressure than the atmosphere. The

atmospheric pressure tends to hold the relatively light ball down in the throat of the funnel.

As a further refinement, a water manometer tapped into the funnel near the point where the ball is closest to the funnel will show a pressure difference of one or two millimeters of water. The ball will float in the funnel with some oscillation and will perhaps rotate on some random axis while maintaining a small annular gap with the funnel.

The operation of Bernouillis's principle is shown somewhat schmatically in the illustrations of Fig. 7-6. If we consider that the wing is stationary and the air is flowing from left to right, we observe in the top illustration that the air on the top of the wing takes a more curved and therefore longer path to get from the front to the rear of the wing. It follows that its velocity must be higher and its pressure lower than the air on the bottom of the wing.

In the lower illustration, we observe that increasing the tilt of the airfoil with respect to the airstream will increase the lift (up to a point) by making the disparity between the upper and lower paths greater. Patterns such as these may actually be seen in a wind tunnel by releasing streams of smoke from a pipe with regularly spaced holes. The streams make the airflow visible.

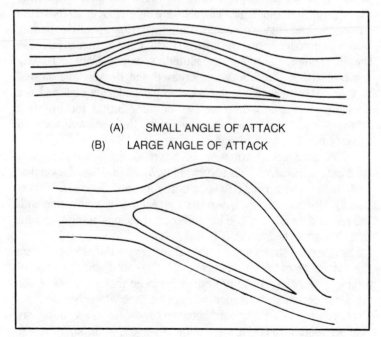

(A)   SMALL ANGLE OF ATTACK
(B)   LARGE ANGLE OF ATTACK

Fig. 7-6. Airflow around a wing.

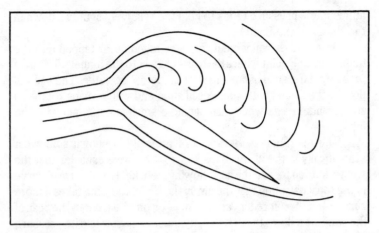

Fig. 7-7. The stalled wing.

Beyond a certain limiting angle, the flow of air above the surface of the wing ceases to be smooth and orderly or "laminar" and becomes disordered or "turbulent." At this point, the wing loses most or all of its lift and the airfoil is said to have "stalled." Figure 5-7 illustrates this situation. This property used to be widely employed in landing airplanes equipped with a tailwheel. The pilot would bring the airplane down so that its wheels were only a foot or so above the runway with no engine power. As the airplane lost speed due to drag he would gradually bring the nose up further and further to increase the lift enough to just hold the altitude constant. With increasing angle of attack the drag also increased and the airplane slowed further. At some speed, the wing would suddenly stall and the airplane would gently drop the two or three inches for a perfect "three-point" landing, coming down with the minimum forward velocity possible for the plane.

The pressure distribution for a typical airfoil operating at a moderate angle of attack is shown in Fig. 7-8. Note that the vertical scale is reversed, reading minus in the upward direction. This is usually done to keep the low pressure readings in association with the top of the wing and the high pressure readings associated with the bottom of the wing.

Figure 7-9 shows some of the parameters of the airfoil including the definitions of the airfoil thickness t, the chord, and the angle of attack. The difference in pressures between the two sides of the airfoil are balanced about some point called the *center of pressure, cp*. This pressure is actually normal to the chord. However, a force is also attempting to resist the motion of the airfoil through the air. Therefore, the net force on the airfoil leans back somewhat from the

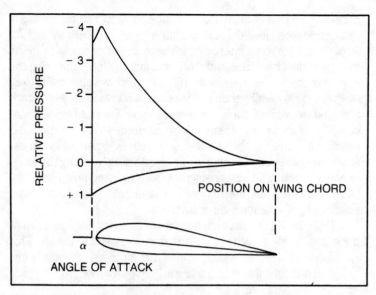

Fig. 7-8. Pressure distribution on the wing.

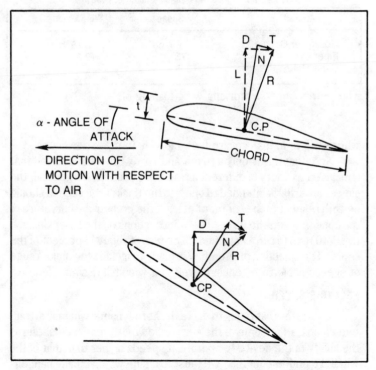

Fig. 7-9. Forces acting on the wing and the effect of angle of attack.

normal N. The resultant force is labeled R in the figure and a retarding force labeled T can be inferred by resolving N and R. Actually we are more interested in the forces at right angles to the motion of the airfoil designated *lift* or L and the force opposing the motion of the wing in the direction of its travel, which is called *drag* D. From the lower illustration we see that a tilt of the angle of attack also tends to rotate R through the same angle. On a real airfoil R also changes its angle with respect to the chord. However the most significant feature is the fact that D increases dramatically. In essence, we have rotated some of our lift into drag by tilting the wing. In actual practice lift actually increases up to some point, but drag increases faster and the L/D ratio is generally greatest at some small angle of attack for thick airfoils.

Both the lift and drag coefficients of an airfoil are a function of the square of the velocity of the airfoil with respect to the air. This speed determines the amount of load that an airfoil can handily lift. This is usually given as wing loading in lbs/square foot. Some typical working numbers are:

Table 7-2. Speed Determines Amount of Load.

Aircraft	Speed	Wing Loading
Gossomer Condor	8 mph	0.15 lb/ft$_2$
Piper Cub	75	12.5
Curtis P-40	225	120

This works out to a working wing loading of:

$$WL = \left(\frac{V}{21}\right)^2$$

where WL is in lbs/ square foot and V is in miles per hour.

Note that this is only a rough-and-ready rule of thumb and that it included a variety of different airfoil cross sections. In general, the slower an airfoil is intended to operate the thicker the section should be with respect to the chord, or at least the greater the curvature of the upper section should be. The wing sections on the Piper Cub and the P-40 tend to run a maximum thickness of about 14 percent of the chord. The selection of aircraft was made to include the major range of speeds liable to be encountered in a windmill design.

RELATIVE SPEED

The factor that tends to make the aerodynamic windmill attractive compared to some of the aerostatic types mentioned earlier is the ability of the device to operate at speeds higher than that of the wind. To understand this, we must examine what actually happens from the viewpoint of some piece of the blade of a large propeller.

Figure 7-10 shows a section of a windmill propeller. The velocity of the wind is represented as an arrow and the rotational velocity of the blade is represented as a second arrow. Because the blade is approaching a given element of air from right angles, and the element of air is also approaching, the two velocities add in the triangular form shown and the wind appears to the blade to be approaching at some small angle of attack. The resultant thrust on the airfoil is tilted slightly backward from the direction of travel. However, it still has a significant component in the direction the propeller blade is moving mechanically. We designate this force MT for *motor torque*. There is also a significant component or R crosswise to MT which we designate EL to stand for the *end loading* that must be born by the thrust bearings on the shaft and the tower itself.

The component MT is the portion of the work which can be presented to the windmill shaft to do useful work. Now let us suppose that for some reason the resistance of the output work were to decrease. In this case, MT would accelerate the propeller and increase V_{blade}. In this case, it may be seen that the wind triangle

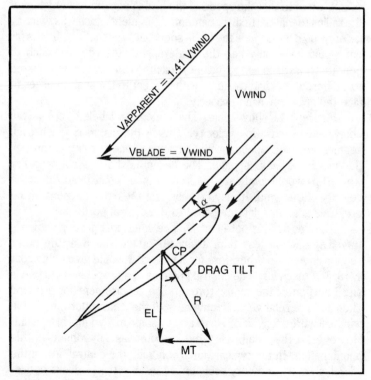

Fig. 7-10. The apparent wind on a blade section.

would flatten and $V_{apparent}$ would swing around slightly so that the angle of attack, α, decreases. Under a no-load condition the speed would actually increase until (α) became very small or even went negative and R would shrink to an equilibrium level, just large enough to make up friction losses.

On the other side of the coin, if the load suddenly increased, this would slow the blade slightly, thereby increasing the effective value of (α) and increasing the magnitude of R. The component MT would then rise, and the machine would find a new equilibrium delivering more torque at a slightly slower speed.

If the torque load were to slow the blade enough, the angle α would exceed the stall angle for the blade and the machine would quickly stop. This stall property is more severe on very high performance machines.

HELICAL PITCH

Having seen that there is a critical relationship between the angle of the blade and the speed of the blade with respect to wind speed, we note that the speed of the blade segment is a function of the radius to that particular segment. A segment which was twice as far out would have twice the blade speed and one that was half as far out would have only half the blade speed. In order for each to operate with reasonable efficiency, each would have to be adjusted to a different mechanical angle with respect to the shaft in order to face the apparent wind properly.

Figure 7-11 shows this. The dimension labeled *pitch* is the equivalent of our wind velocity. This is the distance which each segment of the propeller would advance in one revolution through the air without any slippage. The design is shown for a propeller whose tips would rotate at a theoretical speed equal to six times the velocity of the wind. It may be seen that the outer portions of the propeller are very flat compared to the central portions.

Another point is noteworthy. The innermost portion shown is travelling slightly less than a quarter of the speed of the outer section shown, and therefore for equal area, would give only about 1/18.5 as much lift. The central portions of a propeller yield only a small portion of the motor torque. The requirement for a helical pitch and the relative ineffectiveness of the central portions of the windmill were recognized very early by some millwrights in Holland. However, other countries did not necessarily observe this phenomenon. In the typical Dutch windmill, the central fifth of the wheel is empty and very pronounced twist in the blades is noticeable. French windmills with perfectly flat blades were relatively

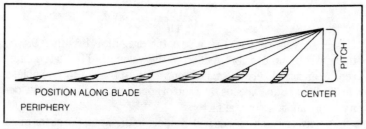

Fig. 7-11. The requirement for helical pitch.

common. While a mill will work with flat blades, the efficiency is much lower, and the machine will deliver considerably less torque at any given rotation speed since some portion of the blade is always stalled.

As one goes further and further out on the propeller, the flatter angle of approach to the wind makes the achievement of a very good lift/drag ratio more and more important. This is shown in Fig. 7-12. Here the blade is shown with a velocity six times the velocity of the wind. Compared with the section shown in Fig. 7-10, we would expect the total force R on the blade section to be:

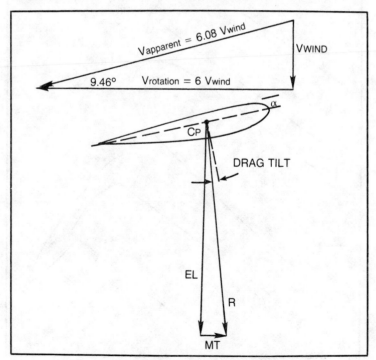

Fig. 7-12. The lift to drag ratio effects on outer blade portions.

$$\frac{F_{outer}}{F_{inner}} = \left(\frac{6.08}{1.41}\right)^3 = 80.29$$

However, you will note that unless the drag tilt is less than 9.46°, there will be no net contribution to the value of MT. Stated differently, an airfoil section cannot contribute to the net torque unless the lift/drag ratio exceeds the ratio of apparent velocity/wind velocity. This implies several factors:

•For very high speeds with respect to the wind, the airfoil must be a basically low-drag type.

•The airfoil must be operating near the angle of attack which yields maximum L/D.

•The airfoil must be very smooth to minimize parasitic drag.

While certain airfoils can yield lift/drag ratios in excess of 30, the state-of-the-art in propeller design seems to limit maximum no-load speeds for windmills to about 11 times wind velocity and maximum power is usually obtained at about 6 times wind velocity.

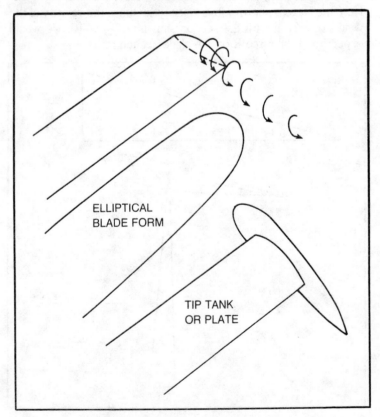

ELLIPTICAL
BLADE FORM

TIP TANK
OR PLATE

Fig. 7-13. Blade tip effects and vortex.

TIP VORTEX

At the tip of the blade, a peculiar condition occurs in that there is a low pressure on the one side of the blade and a high pressure on the other. This gives rise to a tendency for the high pressure air to wash around the end of the blade to create a vortex which trails the tip. This adds to the drag and tends to spoil the L/D ratio of at least the outer portion of the blade. This effect may be minimized by:

•The addition of tip plates or tip tanks.

•Design of the blade to have an elliptical cross section.

•The use of very large diameter to blade-width ratios.

Figure 7-13 shows the phenomenon along with several of the proposed solutions.

At relatively high speeds, the addition of judiciously designed tip plates or tip tanks can considerably reduce the drag. It was discovered during WW II that the addition of tip tanks to fighter planes could often raise the cruising speed by four or five mph. On the propeller, however, the addition of the mass of the tip plate or tank increases the problems of centrifugal force. For many years the wings and propellers of high performance aircraft were designed to present an elliptical form to minimize this effect.

For high-efficiency soaring planes, the wings were simply built with a very high aspect ratio. Gliders with very long narrow tapering wings seem to perform much better than those with short stubby wings because of the tip vortex effect. In essence, since the vortex affects only the region near the tip, this is made the smallest possible percentage of the wing.

On the basis of this very brief discussion of aerodynamics, we shall next discuss the expected performance of a variety of sail and propeller type windmills.

8

Windmill Types and Performance

In this chapter we shall examine a few of the windmill types with an eye to exploring their relative advantages, disadvantages, and limitations. We shall proceed in roughly chronological order.

THE PORTUGUESE OR JIB MILL

Figure 8-1 shows the Portuguese or Jib mill in front and side views. In general, these types which today may be seen in operation in Portugal and on the island of Crete, consist of a series of six masts mounted radially about the windshaft and carrying six jib-type sails. The sail is very similar to the jib of a small sailboat and is sheeted to the tip of the adjacent mast. Two sets of stays are used; one set ties the tips of the masts together, and the second braces the mast against the force of the wind from the shaft which juts forward of the masts. At the hub, opposing pairs of masts are mortised to take the windshaft and lashed together. This provides a crude but sturdy mechanism for transmitting the torque moment from the masts to the windshaft.

As may be seen from the illustration, the chief advantage of this type of mill stems from its very crudity. It can be built by nearly any competent carpenter with a saw, chisel, hammer, and an adequate supply of wood, cloth, and rope.

This mill also has the advantage that it can easily be made large enough to produce significant amounts of power in very light winds. It turns slowly with the tip speed seldom exceeding the wind speed because of the relatively poor airfoil shape blown into the sails.

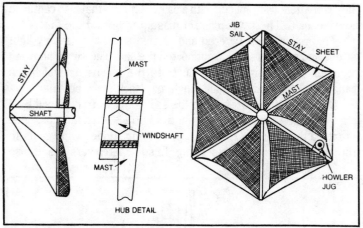

Fig. 8-1. The Portuguese or Jib mill.

The chief disadvantage of this type of mill is that the sails require rather precise adjustment to compensate for changes in the wind. Much like a sailboat, on changing wind the jibs require more or less continuous attention, either taking up the sheets or paying them out as the wind force changes. To alert the miller to the situation, the masts were frequently festooned with large mouthed jugs facing the wind. As the tone of the howling of the jugs changes, the miller is alerted to the requirement to brake the wheel to a halt, go out and change the bottom sheet, allow the wheel to turn a sixth of a turn, and repeat the process until all six sails are reset. In higher winds, the sails are reefed progressively.

It is possible that a mill of this type could be built using roller-reefing gear and some type of roller-sheeting gear operating through a hollow windshaft; however, this takes the machine out of the class of simple machines. With such attachments, however, the machine could be expected to achieve tip speeds equal to or exceeding wind velocity. For light-air locations, there is probably no cheaper wind engine.

THE DUTCH WINDMILL

Figure 8-2 is intended to depict a basic compendium of the Dutch windmill. It is really not very fair to summarize the wide variety of exquisitely designed and crafted machines developed in Holland over a number of centuries with a single drawing and a single figure. However, the drawing does show at least some of the cardinal features of these machines. In general, the opposing masts were square and the same butt thickness as the square windshaft.

They were generally spliced with a pair of timbers that also captured the windshaft. The second set of masts ran behind the first.

The masts were tapered and carried a set of relatively light battens and stringers which supported the sail. The views showing the inner and outer sections illustrate the means by which the change in airfoil thickness and angle of attack were obtained.

On many of the later mills, the sail cloths were replaced by a series of spring-loaded, hinged shutters attached to the ribs. This permitted the shutters to swing open during sudden wind gusts to protect the mill from overspeeding. In some later versions, it also

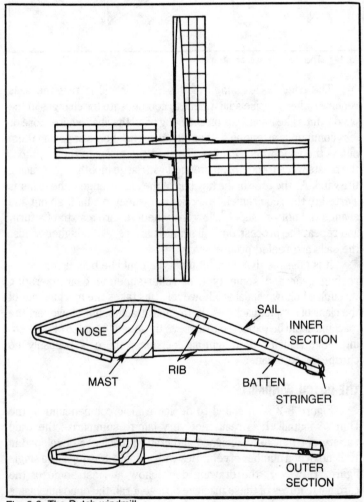

Fig. 8-2. The Dutch windmill.

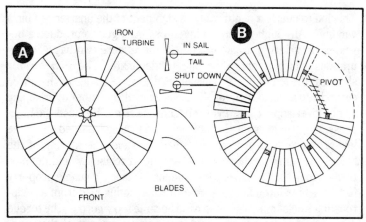

Fig. 8-3. The American wind turbine. The shutter type with one segment shown blown down "out of sail" is at B.

removed the requirement to stop the mill in four different positions so that the miller could furl the sails.

The Dutch windmill would attain no-load tip speeds of four times wind velocity and maximum power at about two times wind velocity. The medium width-to-diameter ratio on the order of six contributed significant tip vortex losses. The open rib framework and general aerodynamic clutter of the ribs, mast, and battens tended to limit the speed and therefore the output of the machines. However, given the state of the knowledge of aerodynamics at the time when these machines were built, the remarkable feature is the use of helical pitch, the tapering airfoil section, and the achieved cleanliness and performance. Apparently without any formal science of aerodynamics, the Dutch millwrights learned by trial-and-error most of the basic parameters of windmill design.

THE AMERICAN WIND TURBINE

From the very earliest settling of the United States, there had been a large number of windmills of the Dutch type used for grain grinding and other industrial purposes. However, with the beginning of the settlement of the plains states there came a demand for a new type of wind machine to pump water from the deep wells required in the dry plains. More significantly, a new requirement arose in the fact that the machines must be capable of unattended operation. Whereas the miller or the blacksmith would be at hand to swing the windmill into the wind or furl the sails when the wind got too high, the farmer would like to have his stock watered even when he was in a distant field and not available to care for the windmill.

This led to nearly a century of development of the unattended farm windmill. Although American inventors managed to produce a bewildering variety of these machines, there were two basic configurations which eventually came to dominate. The *iron turbine* shown at the left of Fig. 8-3 usually had 12 to 18 blades of sheet iron curved into an airfoil section. The blades could be either overlapping or non-overlapping when viewed from the front. Because the high-pressure stream from the bottom of one blade passed over the low-pressure stream from the adjacent blade, the stalling of the blade was significantly retarded, and the machines would develop tremendous torque at very low speeds. This is a valuable property for a machine intended to handle the alternating load from a reciprocating well pump. When the wind became too vigorous, the torque developed by the wheel would swing the wheel out of the wind, parallel to the flat rudder or tail that normally kept it pointed into the wind.

The *shutter* type had a large number of segments of very much smaller blades placed closer together. The segments were usually spaced with a few "missing teeth" for clearance. In this type, the segments would blow down roughly parallel with the wind when the wind became too strong. The blades or entire blade fans on these units were often made of light wood therefore fewer of these survive. The shutter type had an advantage in lighter air because of the lightness of the blades.

The iron turbine was capable of attaining a tip speed of something like 1.75 times wind speed and yielded maximum torque at something like 3/4 wind speed. Windmills of this type are still available commercially from a few sources that sell them primarily for water pumping.

POWER EXTRACTION

All windmills obtain power from the wind by slightly slowing the wind going through the wheel. Obviously, the wind cannot be brought to a dead stop behind the wheel since this would cause a "pileup" of air behind the wheel which would stop the propeller. It can be shown mathematically that no windmill can ever extract more than 59.2 percent of the power from the air column. Windmills are therefore rated on the basis of a power coefficient (P_c) which is the fraction of the available power extracted. The curve of Fig. 8-4 shows the comparison between the efficiency of the various types. It should be noted that the American wind turbine is a relatively modern machine of comparative aerodynamic sophistication compared to the Dutch type represented. With comparable

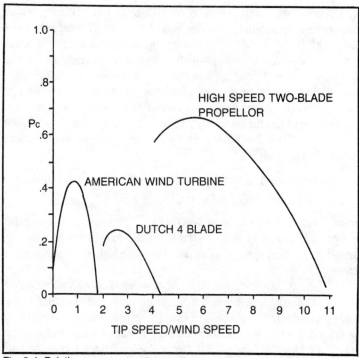

Fig. 8-4. Relative power extraction coefficients from wind tunnel tests.

aerodynamic cleanliness, the Dutch machine would rise vertically and to the right.

The unmistakable conclusion from the data, however, is the fact that the nod for overall efficiency must go to the high-speed, two-blade type machine. In the recent surge of work in energy-related areas, it is only the high-speed machines of the propeller and Darieus types which have received significant attention.

HOW MUCH POWER CAN YOU GET?

The answer to this question gets a bit more sticky. In general, the power derived from a wind engine is a function of the aerodynamic cleanliness, the tip speed/wind speed ratio for maximum power, the square of the radius of the rotor, and the cube of wind velocity. Unfortunately, the reported powers from windmills are nearly always given at the highest wind encountered when the windmill was tested. Depending upon the location, this may have been a 10- or a 35-mph wind. In one case, a 70-mph wind was used for the machine rating. The curve of Fig. 8-5 was made up from data from various sources. The data has all been normalized by the cube-of-wind-velocity ratio to a 15-mph wind velocity which would correspond to an energy wind for about 75 percent of the country. The curve includes some relatively small machines along with the largest and most powerful windmill ever built. The Table 7-1 may be used to extrapolate the data to other wind velocities.

Going back to our energy wind concept, if we presume that 70 percent of our energy would be produced by energy winds blowing 42 percent of the time and we take our energy wind as 15 mph, we may look at the 2-kW point and see that a 30-foot *very efficient* windmill would give us:

$$2000 \text{ W} \times 1/0.7 \times 0.42 \times 24 \text{ hrs/day} \times 30.44 \text{ days/mo.}$$
$$= 876,600 \text{ watt hrs/month}$$

or an average power drain of 1200 watts. If this energy is twice the energy for a wind of average velocity, the average wind velocity would be 13 mph, which is available only in the high-wind areas of the country.

It is also noteworthy that if energy winds blow only about three out of seven days, we would have to be prepared to either store rather considerable amounts of energy or to content ourselves to operate the high-drain appliances only at times when the energy winds are blowing. The 1.2 kW figure does not track too badly with our average household drain computered in Chapter 1 if we presume that the home electrifier will have developed some sense of conservation by the time he saw how much effort is required to extract power from the wind.

The strength and frequency of the winds is obviously a very central issue in the selection of windmill types. It is generally considered that an 8-mph yearly average represents the minimum wind velocity practical for a two-blade propeller type. Below this speed some multiblade type has the clear advantage. It has been shown that a very light multiblade wheel, which must be swung out of the wind at 15 mph and operates from 6 to 15 mph, will produce a monthly average power 14 percent greater than a heavier engine which operates only from 8 to 15 mph.

A small multiblade type windmill, similar to the farm types used for water pumping, is generally not capable of driving a modern high-power automotive alternator to full output in winds less than 20 mph as seen in the curve of Fig. 8-5, combined with the multiplier of 2.37 from Table 7-1. In the light-wind areas which make up a significant fraction of the country, a very large windmill is required in order to come up to the 1 kilowatt level monthly average. Something on the order of a 50-foot Portuguese wheel with reefing and sheet-

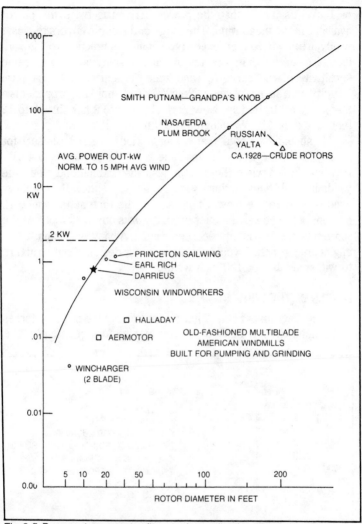

Fig. 8-5. Power out versus rotor diameter (normalized to 15 mph average wind).

ing handled by electric motors under automatic control would probably do the job, but the expense for a structure of this size would be very large and the design of a safe, stable and efficient structure would be a major assignment in mechanical engineering. It is highly questionable whether the development and construction of a wind-electric plant for the home can be justified for an area where the mean-annual wind is less than 8 mph.

The curve of Fig. 8-6 gives the diameter of propeller required for a two-blade propeller-type windmill producing an annual average

129

of 1 kW, assuming that the power derived is twice the power available in the mean wind. The range of the curve covers the least windy areas where a propeller type engine is practical to the windiest area in the United States, Mount Washington. The extreme sensitivity of the curve to wind speed is apparent. In a typical "windy" area such as Albany, New York (9.0 mph) the propeller is a 25-foot giant. In Buffalo, New York (14.6 mph) it has shrunk to 12 feet, and atop Mt. Washington (36.9 mph) it is a 3-foot pigmy!

It should be noted that 39-mph winds are not unusual for Albany. At these times the machine would be producing 3.4 kW. However, days when the wind is less than 5 mph and when the windmill would hang stationary are not unusual either. If the energy wind is assumed to blow 42 percent of the time and produce 70 percent of the power the energy wind works out to 39 mph. At the same time the maximum recorded wind in Albany is 71 mph. Good engineering practice would probably call for design of the structure to withstand 100 or 120 mph winds.

FORCES OF THE WINDMILL

The diagram of Fig. 8-7 illustrates some of the principal forces acting upon the windmill. If we consider a small segment of the

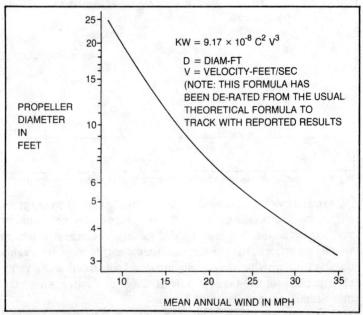

Fig. 8-6. Diameter of a two-blade propeller required to produce an annual average power of 1 kW in the mean annual winds shown.

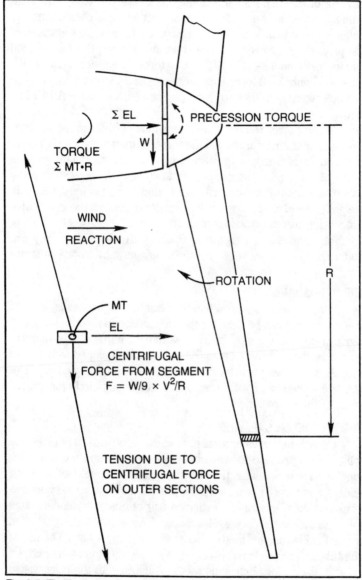

Fig. 8-7. The forces acting upon a windmill.

blade, we see that it is acted upon by the motor torque (MT) and the end-loading force (EL), both of which are due to the aerodynamic performance of the blade. It is also acted upon by centrifugal force due to its own weight, velocity, and radius of rotation and it must sustain the action of centrifugal force due to the portions of the blade outside of itself. By slightly tilting backward the blade, the centrifugal force can be made to counteract the EL component and reduce the bending moment on the blade. If the rotational speed of the propeller is allowed to follow the wind velocity, this cancellation cannot be complete since the EL force is proportional to V^3, whereas centrifugal force is proportional to V^2. Advantage is taken of this property on modern wind engines such as the NASA Plum Brook unit.

At the power the thrust bearings must withstand the summation of all of the EL forces and the torque summation of all $MT \times R$ torques. In addition, if the windmill is allowed to rapidly slew in gusts, there is a gyroscopic precession torque in the vertical plane which acts upon the windshaft. The windshaft also must bear the weight of the propeller assembly. In order to restrict the gyroscopic precession forces to safe limits, large modern wind engines are generally not allowed to swing freely with the wind. They are generally moved by a servo-control loop operating from a separate wind direction sensor.

THE WINDSHAFT

For the home construction of medium size windmills, a large truck rear axle can be used for the windshaft and pivot bearings. A modern truck axle has the bearings within the hub and is designed to withstand large eccentric loads. It also comes equipped with a large and effective brake which can be employed with the windmill. The hub is also equipped with a large and sturdy bolt circle for attachment of the propellor.

THE PRINCETON SAILWING

One of the developments in modern aeronautrical engineering which gives promise for an economical construction technique for medium size windmills is the sailwing technique developed for hang gliders and ultra-light aircraft. Figure 8-8 shows the very simple construction of this airfoil section. A pair of ribs and a leading edge fairing are mounted upon a tubular spar. The fabric covering is pulled around the fairing and held at the trailing edge by a wire. The use of the tubular spar is especially handy for fine adjustment in pitch. Of course, the major pitch adjustment is built in with the relationship between the angles of attack of the inner and outer ribs. This feature would also be handy for feathering of the rotor during high winds which would exceed the capacity of the machine.

VARIABLE PITCH AND FEATHERING

The addition of a variable pitch mechanism is a rather expensive feature to add to a wind engine. It is attractive in that it extends

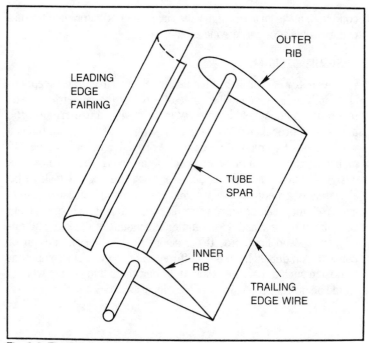

Fig. 8-8. The Princeton sailwing construction.

the range of wind speeds over which the machine will operate effectively. However, the entire pitch curve of the helix would have to be variable to be truly effective. In general, this is not done because a mechanism for changing the twist of the blade would be difficult to produce. With an airfoil such as the sailwing, it is a possibility since it would be necessary to rotate only the outer and inner ribs independently. But this would entail some stretching of the trailing edge wire and the fabric. If an automotive type alternator is used, the maintenance of a constant speed is not required, and the advantage accruing to the use of variable pitch is limited.

On the other hand, in the matter of stowage during storms, the ability to feather the propeller would be a distinct advantage. When the propeller is stationary, the wind resistance of the blades is about equivalent to that of a flat plate of the same area and will increase as the square of the wind velocity. If the helical blade can be turned to the most favorable angle, the wind resistance of the blade can be reduced to about a fifth of the resistance in the normal pitch condition. Since the bending moment of the blade at the hub is approximately proportional to the cube of wind velocity and the propeller does not have the advantage of centrifugal force in the stationary condition, blade failure in high winds in the locked rotor condition is a major problem on large wind engines.

WINDMILL DESIGN

As noted earlier in the chapter, unless you live in an exceptionally windy area, a windmill capable of providing the electrical needs of a modern household is a pretty large device. Furthermore, the high speed propeller must be pretty sophisticated. At six times a wind speed of 40 mph, the tips of the propeller will be turning 240 mph, and the forces on the tower will be measured in tons. The design of such machines is definitely not the sort of task to be allocated to someone not thoroughly versed in mechanical engineering. A 25-foot diameter propeller disintegrating at high speed from centrifugal force could do an immense amount of damage and endanger life and limb. For this reason, no "cookbook" design or construction details have been offered here. Engineers competent to design such a machine would not need them and people who do would be better off obtaining professional assistance in the matter.

Water Pump

The watermill is—if anything—older than the windmill. An epigram by Antipater of Thessalonica (ca. 85 BC) describes a mill operated by water. Vitruvius describes the Roman mill with a horizontal axle, a geartrain, and a millstone set in De Architectura written about 15 BC. The Chinese record the use of a horizontal water wheel to operate a bellows for a blacksmith's forge in 31 BC. For those interested in a historical and artistic background of the watermill—and the windmill as well—I would recommend *Windmills and Watermills* by John Reynolds, published by Praeger Publications, New York. From these early beginnings, men have extracted energy from falling water. Unlike the windmill, the use of water power continues on a large scale to the present day, although the rustic mill has been replaced by the hydroelectric power plant.

The vertical axis, horizontal turning water wheel is usually considered to be older than the horizontal axis vertical turning wheel. The former was usually termed a "Greek" wheel while the horizontal axis machine was termed a "Roman" wheel. Surprisingly, the modern hydroelectric plant is more closely related to the older "Greek" design in the majority of cases. The reasons for this will be discussed shortly.

HEAD AND FLOW

The energy available from water in the normal sense (not including thermonuclear fusion) is determined by the distance the water can be allowed to fall through the machinery, or the *head*, and the amount of water that is available for processing through the machine, called the *flow*. Significant amounts of energy can be extracted from a small amount of water falling a great distance or from a very large amount of water falling a small distance. The machinery differs for the efficient extraction of the energy.

An example is shown in Fig. 9-1, labeled the Greek Turbine. In this unit, the water was allowed to fall down a chute in order to exchange some of its potential energy (due to height) for kinetic energy (due to velocity). At the end of its fall, it impinges upon the blade or blades on one side of a wheel fitted with nearly flat blades. The force or reaction pushed the blade around, and the water drained out below.

There are some obvious losses with this type of unit. First of all, the exchange of potential energy for kinetic energy is not terribly efficient since some of the energy is lost in turbulence and friction in the chute. Secondly, some of the kinetic energy is wasted in splashing off of the blades. Thirdly, the blades can never absorb all of the kinetic energy since this would bring the water to a full stop and flood the unit. Finally, the unit must give away some head to allow the water someplace to fall.

The Irish Turbine shown in Fig. 9-2 can alleviate a few of these problems. It receives the water on a nearly horizontal trajectory. The curved shape of the blades will slow the water to a near stop and eject it horizontally. Since the flow is smoother, it can extract a

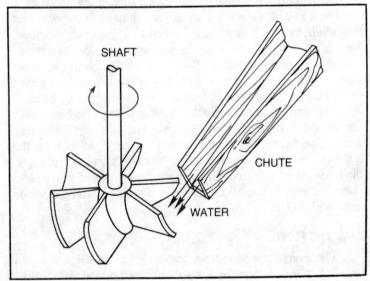

Fig. 9-1. The Greek turbine.

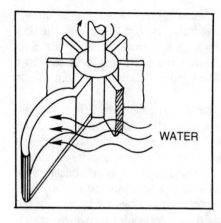

Fig. 9-2. The Irish turbine.

higher percentage of the kinetic energy. It also gives up less head to clear the water.

A somewhat more modern approach is shown in Fig. 9-3. In this machine, an outboard motor lower unit—complete with gearbox for right angle drive—and a venturi inlet pipe is used. The casing

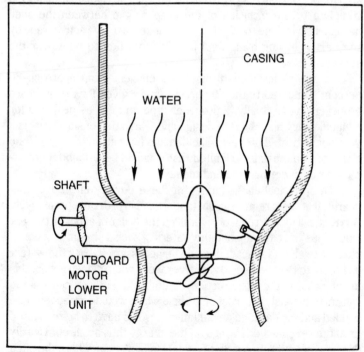

Fig. 9-3. A machine using an outboard motor lower unit with a gearbox for right angle drive and a venturi inlet pipe.

is kept deliberately large to minimize friction losses. Just before the propeller, the tube necks down into a venturi to increase the flow speed before being forced through the propeller. A typical 5-hp lower unit will provide an output on the order of 3-hp at 3600 rpm when operating under an 11-foot head and discharging about 252 lbs of water per second.

There are several attractive features about this. First of all, the power is on the right order for a home electrical installation with the 3 hp being on the order of 2.2 kW and secondly, the speed is appropriate for driving a 60-Hz alternator without gearing. Also, the outboard lower unit is generally available at a very reasonable price if purchased used on an otherwise defunct outboard motor.

The real class in this department belongs, of course, to the various impulse wheels designed specifically for hydro power installations.

Figure 9-4 shows the cross section of a Pelton wheel bucket. This unit in three dimensions looks very much like the fully opened shell of a clam. The stream of water from the nozzle is split into two, and the curvature of the buckets is such as to extract the maximum amount of available energy from the water. The clamshell analogy is complete to the inclusion of the cusp or gap between the shell halves. This is done to minimize the blanketing of the active blade by the next following blade until it is in a position to take over the stream.

In a well-designed turbine, the efficiency can approach 80 percent compared to the 60 percent figure given for the outboard propeller unit. Actually, a propeller type unit can be designed for high efficiency also, but the design and fabrication of such units is a very sophisticated process. The demand for units in the 3- to 10-hp class for home use is so limited that it would be a doubtful investment in engineering time and money.

The principal characteristic of these devices is the fact that nearly all of them require a substantial head in order to operate well. A typical fall for a stream or river is on the order of a foot or two per mile. If we were to impound the waters of such a stream in order to obtain a twelve-foot head, it would be necessary to back the waters up for six to twelve miles. If the river were not in a deep gorge, this might require the construction of embankments to contain the waters. Obviously, the cost of the land which would consequently be flooded and the dam and embankments would have to be recovered from the revenues or savings on the energy thus acquired. Usually this type of installation is justified only in hilly areas where the pitch of the stream is great and where the stream is contained within a

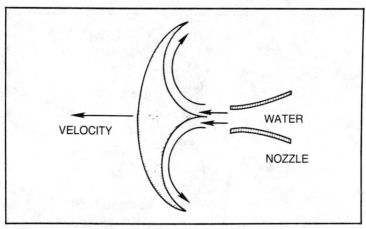

Fig. 9-4. The Pelton wheel bucket in cross section.

natural gorge. Since this represents something of a special case in geography, it follows that such medium head installations are not very common.

ENERGY IN FALLING WATER

Before going further, it might be instructive to consider the amount of energy which can be obtained from falling water. We shall start by considering what must be about the simplest water engine possible. Figure 9-5 shows a bucket attached to a rope wound around a drum. The bucket contains one cubic foot of water which—in the case of fresh water—weighs 62.4 pounds. The bottom of the bucket is 10 feet above the ground. The bucket has a potential energy of:

$$62.4 \text{ lbs} \times 10 \text{ feet} = 624 \text{ foot-pounds}$$

If friction is neglected, this energy could be delivered through the rotating shaft to an external load.

If the external load were to constrain the descent of the bucket such that it descended the 10-foot head in 10 seconds, the output rate would be:

$$\frac{624 \text{ ft-lb}}{10 \text{ sec}} = 62.4 \text{ ft-lb/sec.}$$

This is equivalent to:

$$\frac{62.4 \text{ ft-lb/sec}}{550 \text{ ft-lb/ sec/ horsepower}} = 0.113 \text{ horsepower.}$$

This is in turn equivalent to 84.6 watts. Our more common unit in electrical energy is the kilowatt-hour (kWh), and 1 kWh = 2.656×10^6 ft-lb.

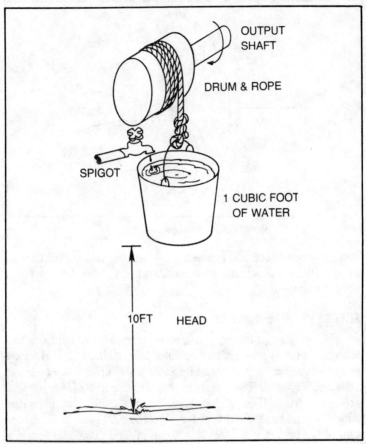

Fig. 9-5. The simplest water engine.

Next, let us imagine that we are able to use our 10-foot head in some perfect, lossless machine to generate electricity on a continuous basis. Since, at the 10-foot head each cubic foot has a potential energy of 624 ft-lbs we would require:

$$\frac{2.656 \times 10^6 \text{ ft-lb/ kWh}}{624 \text{ ft-lb/ ft}^3} = 4256 \text{ ft}^3/ \text{kWh (at 10-foot head)}$$

With a one-foot head we would require 42,560 cubic feet, and with a 100-foot head we would require only 425.6 cubic feet.

If the water were coming from a pond and we were to require that the drainage would lower the pond by no more than one inch, the pond would require an area of:

$$\frac{4265 \text{ ft}^3}{1/12 \text{ ft}} = 51,072 \text{ ft}^2 \text{ (at 10 foot head)}$$

which is:

$$\frac{51,072 \text{ ft}^2}{43,560 \text{ ft}^2 \text{ per acre}} = 1.17 \text{ acres}$$

For the one-foot head we would require 11.72 acres, and for the 100-foot head we would require 0.117 acres.

Now we would probably like to replenish our water supply by means of rainfall, and the rain does not fall evenly in all places. On a national average basis, the rainfall can be as little as three inches per year in Egypt and Arabia and as much as 119 inches in Samoa and the Solomon Islands. However, the rainfall in Rochester, New York averages 36 inches per year so let us consider that figure. Because there are 8,766 hours per year the 36-inch rainfull works out to an average of 4.11×10^{-3} inches per hour or 3.42×10^{-4} feet per hour. To obtain our 4,256 ft^3 of water, we would have to drain 1.244×10^7 ft^2 or 285.5 acres or 0.835 square miles. (There are 640 acres per square mile). And all of this would provide only an average of one kilowatt provided that we could have a 10-foot head.

The message in this little study is simply that there is more than one house per every 0.8 square mile in the area of Rochester, New York, which has an above average rainfall. We cannot all obtain our electricity from the rainfall. In matter of fact, the Rochester Gas and Electric Co. does have a hydro plant on the Genesee River in the heart of Rochester. The river is not small and the company is blessed with an 85-foot waterfall. Roughly half of the river goes through their turbines. However, the hydro installation, which is on-line more than 98 percent of the time, manages to produce only about 5 percent of the electricity used in the area.

LOSSES

In actual practice, only a fraction of the energy from rainfall is available. First of all, the soil absorbs a healthy fraction which is largely given up by plants to the atmosphere. The reservoirs also lose a good bit to evaporation. In addition, there are losses of head which are required for the water to flow to the power station. Harking back to Fig. 9-5, we note that the bucket was filled at the top but that it is the bottom of the bucket which hits the ground first so that of the total head actually available, we lose the height of the bucket. Even if this simple machine had no losses from friction, we could obtain only a part of the available energy.

THE OVERSHOT WHEEL

The overshot wheel is the picture that comes to mind for most people when they hear the term *waterwheel*. However, the traditional overshot wheel was a very inefficient device. As may be seen from Fig. 9-6, the trough filled the buckets at the top. However, as soon as a bucket began to turn from the top position, the water would commence to spill out. By the time that the bucket was level with the axle, the water was entirely gone. As shown on the graph to the right, about half of the water was used for about half of the diameter; thus, the wheel could extract only about 25 percent of the energy.

This loss can be partly eliminated by "feathering" the wheel as shown in Fig. 9-7. The buckets on a feathered wheel are held upright until they near the very bottom of the wheel whereupon a mechanism dumps them. A wheel of this sort must turn very slowly in order to avoid losing the water due to centrifugal force. It must therefore be rather large.

Suppose for example that we wish to construct a wheel to give us 2.2 kW (3 hp) with a 10-foot effective diameter. We would need to handle 4,256 ft³/hr × 2.2 = 9363 ft³/hr. If the wheel turned at 10 rpm or 600 revolutions per hour and the wheel had 35 buckets, the water would be handled at the rate of 600 × 35 = 21,000 buckets per hour and each bucket would have to have a capacity of 0.446 cu ft. If the buckets were semi-circular and 1 foot in diameter, they would have to be about 14 inches wide. The overall diameter of the featured wheel would have to be something like 11 feet to get a 10-foot effective drop and still clear the water on the bottom. This is

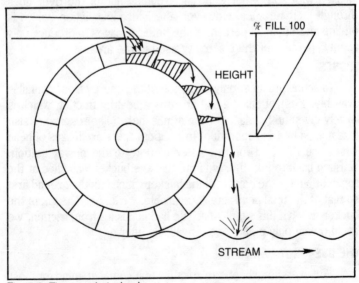

Fig. 9-6. The overshot wheel.

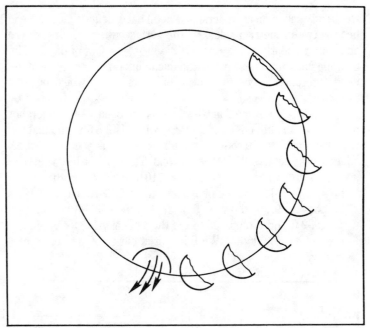

Fig. 9-7. The feathered wheel.

a pretty substantial sized piece of hardware compared to the propeller arrangement with the same rating of Fig. 9-3. When the trough clearance for filling and the bottom clearance are added, a head of about 12 feet would be required in order to operate the unit. In terms of overall efficiency, the unit would use less water to produce the same power than the propeller did. However, it would require gearing up by a factor like 360 to attain sufficient speed for an alternator.

LOW HEAD WATER ENGINES

In Europe and England, the use of the undershot wheel was far more common than the use of the overshot wheel for the simple reason that there are far more sites suitable for a low-head installation than there are for a medium-head installation. It is far easier to obtain a 1-foot head than it is to obtain a 10-foot head. A number of these were located at the mouths of tidal estuaries and ran in one direction when the tide was coming in and the other when it was going out.

Figure 9-8 shows the working portion of an undershot wheel. The wheel is fitted into a close-fitting millrace to minimize losses around the sides and bottom. The one paddle (or float as they are

143

frequently called on an undershot wheel) has a definite difference in head between the two sides. At the surface, the pressure is zero and climbs linearly at the rate of 62.4 lb/ft^2 per foot of depth. The average pressure is one half of the maximum pressure so that if the head difference is one foot the average pressure would be 31.2 lb/ft^2.

Now let us assume that we would like to build a 3-hp undershot mill and that the effective radius of the wheel is 9.5 feet and that the wheel is turning at 10 rpm. At the effective radius point, the wheel speed works out to 9.95 feet per second. The work on the wheel is:

31.2 lb/ft width × 9.95 ft/sec = 310 ft-lb/sec/ft width. For 3 hp = 1650 ft-lb/sec, we obtain a width of 5.32 ft. The mill will be passing 9.95 ft/sec × 5.32 ft × 1 ft × 3600 sec/hr = 190,562 ft^3/hr. Compared to the theoretical value for a 1-foot head of 43,560 ft^3/hr/kWh this works out to 50 percent efficiency.

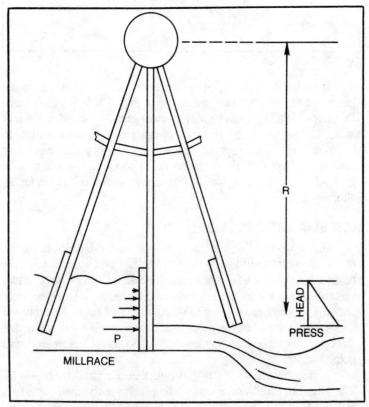

Fig. 9-8. The undershot wheel.

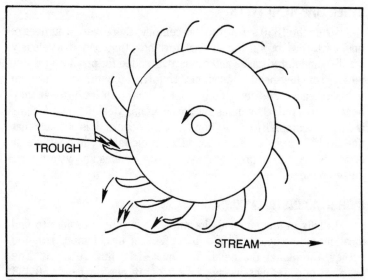

Fig. 9-9. The breast-shot wheel.

In an actual installation, the efficiency would not be this good because leakage around the floats would reduce the effective head and the water passed through would not immediately depart. Therefore, it would cause a certain amount of back pressure on the wheel. A practical 3-horsepower wheel would probably be about 8 feet wide and 20 feet in diameter. This is an awfully large piece of equipment compared to the outboard unit. However, it will recover "free" energy in a very low density situation. In actual practice, it would probably be necessary to have something like a 1-¼ or 1-½ feet of head to allow for the millrace entrance and exit. A wheel such as this will actually operate in an open stream without the fitted millrace, but the power is greatly reduced.

THE BREAST-SHOT WHEEL

For installations where the head is more than a foot but not as much as 10 feet, the breast-shot wheel provides an optimum solution. As may be seen from Fig. 9-9, the breast-shot wheel combines some of the features of the overshot wheel and the undershot wheel. Because of spillage, it gets only about 25 percent efficiency. However, it will operate at medium heads quite effectively. With end-plates (not shown in the drawing), the breast-shot wheel does not require a close-fitting millrace for effective operation. For many years, the breast-shot wheel was the wheel of choice for low head installations.

OTHER LOW-HEAD TYPES

From the 1880's well into this century, there were a number of manufacturers producing special turbines that had marvelously scrolled and fluted cases and runners intended for producing water power from low-head installations. Unfortunately these units seem to have disappeared completely from the marketplace because they became impractical for industrial applications. Most of these designs are far too complex for most home workshoppers. It is possible that one of these units might be found in an old factory in operable or repairable condition provided that the owner does not want it for a museum piece.

HIGH HEAD INSTALLATIONS

In hilly or mountainous locations, it is often possible to find locations where heads of 100 or more feet can be obtained. As noted earlier, the higher the head, the smaller the flow required. The pressure of water increases by 62.4 lb per ft^2 per foot head or 0.433 lb per in^2 per foot head. At a 100-foot head the pressure is 43.3 lb per in^2 which is well within the rating of the oil pumps used on diesel trucks. These pumps are also generally of the positive displacement type and will run equally as a motor. Ratings in the one- to three-horsepower range are not unusual for large diesel engines so that these units may generally be adapted to home electrification usage. They are not generally constructed of corrosion resistant alloys so that pump life may be limited to a few years. If the price is reasonable, though, this may not be a severe drawback. A commercially made water pump of suitable pressure and power rating may also be selected from a number of vendors.

SYSTEM CONSIDERATIONS

A basic hydroelectric system consists of the following items:
- A source of water
- A place to dispose of the water
- A strainer or grating to keep debris out of the system.
- A valve or gate to permit system shutdown.
- A water engine of some sort.
- A generating system to produce the electricity.
- An electrical control system for control of voltage and frequency
- An overflow or bypass mechanism so that heavy rains do not cause flooding.

Starting with the matter of the valve or gate for system shutdown, it is usually advisable to place this at the "high" or "entry" point of the system. When placed at the top, this valve may be shut and the system drained for maintenance. This is particularly required in high head systems, but even on an undershot wheel, it is nice to be able to dry the millrace for maintenance work on the wheel.

In the case of the generating system a choice sometimes exists. When it is easy to regulate the speed of the wheel by means of the inlet gate, it is often possible to directly drive a 60-Hz alternator and accurately control the speed. This is probably the least expensive and most direct approach. In cases where the wheel speed is likely to be widely varying, it is probably easiest to directly drive a DC generator and then use a motor-generator set to obtain the requisite 60 Hz for the final usage. This is expensive because it requires three electrical machines, rather than one. However, it is probably more effective than the use of a variable speed mechanical transmission. If the entire system head is not used at full load, it is usually possible to produce a servo-control for the inlet gate to hold the output speed constant under varying load. This is the approach used in large commercial hydro installations.

The overflow or bypass system will often consist of a wire cut into the top of the impoundment dam with a bypass channel dug to permit excess water to flow around the powerhouse and directly into the tailing stream. In areas where heavy spring rains are the rule, care must be taken in the design of the bypass to ensure that it is adequate to handle the heavy runoff. Also the tailing stream channel must be large enough to handle the runoff so that it doesn't raise so high that it backfills the engine runoff. This backfilling can seriously reduce the output of the machine since the water must be able to fall freely out of the machine to provide a reasonable flow. Of course, an even worse case exists when the tailing stream rises high enough to flood the powerhouse. This can lead to very expensive damage to the electrical components, and the flooding will usually leave the powerhouse filled with mud. Worse still is the possibility that a really severe flood can wash away the whole powerhouse.

The greatest danger of all is the possibility that a series of heavy rains can weaken and perhaps rupture the impoundment or the dam. In this case it is likely that the entire contents of the impoundment will be released in a single tidal wave. The destructive force of a ten foot wall of water rushing down a gorge or stream is almost beyond belief. Bridges, culverts, houses, or nearly anything that gets in the way can be swept aside like matchsticks.

A water impoundment of any significant size should never be erected without the services of a competent civil engineer. Local laws, ordinances and design codes should be rigorously observed and investigated before any such construction is initiated.

A HOME HYDROELECTRIC INSTALLATION

Figure 9-10 shows some of the principal features of a small medium-head home hydroelectric installation. The unit uses a propeller type motor of the sort shown in Fig. 9-3. At the water inlet, a trash rack is placed to keep branches, rocks, and large debris from falling through the motor and damaging the propeller or gears. A belt drive is used to elevate the alternator some distance above the tailing stream to reduce the likelihood of flood damage.

The speed of a propeller type water engine is a function of the torque load and the head. Therefore, a certain amount of head is given away for purposes of control. A frequency sensor measures the frequency of the load current and operates the gate control motor to adjust the head accordingly. If a heavy load comes on line, this will tend to slow the alternator, thereby producing a phase lag or a reduction in frequency. The control responds by opening the control gate which raises the head in the casing, which in turn offsets the lag. Voltage and frequency controls can be as described in the section on motor generators. Depending upon the speed of response of the water level in the casing and the inertia of the motor-generator system, it may or may not be possible to operate the system in a true phase-locked condition.

The compressed air system is used to provide a continuous stream of bubbles for an area directly in front of the inlet. This will tend to circulate warmer water from the bottom of the impoundment to the surface and aids in keeping the inlet ice-free in cold climates. The depth of the impoundment should be somewhat deeper than the greatest depth of ice generally encountered in the area.

It is probably advisable to install the switchboard and control system in a separate room that is well above the water outlet. That area will generally be completely saturated in humidity, and running condensation on the walls and motor casing will be very likely. The presence of so much moisture is undesirable around the electrical equipment. Extreme moisture also raises the likelihood of annoying electrical shocks when the equipment is touched, because it would be nearly impossible to keep the insulators free of moisture.

An installation of this sort, while obviously expensive will provide about the finest type of home electrical service. It has the advantage of being nearly as dependable as the water supply itself and requires no storage of electrical energy. The system can be easily built so that very large overloads can be handled temporarily

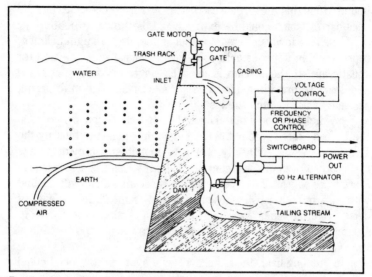

Fig. 9-10. A small hydroelectric installation.

by simply opening the control gate wider during the overload. The requirement for an elaborate prioritizing and load-scheduling system is either eliminated or minimized. In short, if one has the money, geography, and legal feasibility to permit the construction of such an installation, it would represent a premium quality home electric plant.

ENVIRONMENTAL IMPACT

The establishment of a hydroelectric plant always has some significant impact upon the surroundings. In particular, the construction of the impoundment will probably flood a great deal of land. If you do not own this land, it will be necessary to buy it in most cases. It is not unlikely that roads will have to be re-routed and so on. In populated areas this is liable to require a certain amount of political activity.

The people living downstream from the installation are also apt to have a few words to say about the subject. During the initial fill period, the flow of their stream will be either completely shut off or greatly diminished. Once constructed and filled, they may feel that the dam represents a substantial threat to their life and property because of the tremendous damage that a failure could produce.

In terms of ecological impact, there might also be certain repercussions. During the initial fill period, the reduced flow is very likely to damage or destroy the downstream aquatic life. Someone is

149

bound to find that the stream carries an obscure sub-species like the snail-darter which could be extinguished by the construction.

Even the lowly undershot wheel, if operated in a tight millrace, represents an impenetrable obstacle in at least one direction for aquatic life. If the stream is used by migratory fish, it may be necessary to provide a fish ladder or similar installation to permit passage of fish around the installation.

By the way, some of the finest fishing to be found anywhere in the United States is found in the huge impoundments built by the Tennessee Valley Authority. These impoundments provide a huge reservation for game fish where no fish previously existed because there were only shallow rivers. In fact, these impoundments abound with bass and other fish whose ascendency came at the expense of certain riverine types. Whether these changes represent an ecological improvement is probably a matter of viewpoint. It is certainly a subject of debate.

In the long haul, the installation of a hydroelectric plant could provide a nice environment for aquatic life if it is carefully handled. The net flow of the stream will be unaltered. The power is produced with no pollutants and no wastes to dispose. However, the acts of construction cannot but help to at least temporarily alter the environment. Because of the sheer size of the project this may bring vigorous objections from some quarters. There is probably no single answer to the question.

10

Electricity-From-Sunshine Principles

Every reader knows that the earth revolves about the sun at the center of the solar system. However, for the purposes of our discussion here, it is necessary to dig a little bit deeper into our relationship with the sun. As will be noted shortly, the details of this relationship have a significant bearing upon the design of any solar powered electrical system.

The earth revolves about the sun in a nearly round elliptical orbit with the sun at one focus at a mean distance of about 93 million miles. The equator of the earth is inclined to the plane of the ecliptic by an angle of about 23.5°. This relationship can be visualized from the diagram in Fig. 10-1. The direction of the rotation of the earth with respect to the revolution about the sun is such that one revolution is subtracted in each circuit. Stated another way, every 365 days, 6 hours, 9 minutes and 9.54 seconds of mean solar time, any fixed star will have passed the meridian 366 times. This is the reason for the changing of the constellations with the seasons.

The 23.5° inclination of the equator is even more significant in terms of solar energy and climate. Surprisingly, the foci of the ellipse are located so that the earth is actually slightly closer to the sun on the 21st of December than it is on the 21st of June. However, the slight increase in the received radiation is completely over-shadowed by the fact that on December 21, the sun is very low in the sky and we experience the shortest day of the year. The lower part of Fig. 10-1 shows the geometry of this situation. Not only is the day shorter at the high latitudes in the northern hemisphere but the zenith angle stands at its largest value. At 43° North at noon on December 21, the zenith angle is approximately 66.5°. The amount of sunshine falling on a horizontal surface varies as the cosine of the zenith angle and is 0.399 for 66.5°. The sun gets only 23.5° (90° − 66.5°) above the horizon, and it is winter in Rochester, New York.

By contrast, on the June 21, the zenith angle is 19.5° (43° − 23.5°), for which the cosine is 0.943. On a horizontal surface in Rochester at noon, there is nearly three times as much energy incident on the first day of summer as there is on the first day of winter.

For a solar collector, we can beat the cosine law simply by turning the collector so that it is always aimed directly at the sun. However, it may be seen that the distance the sunlight has to travel through the atmosphere is proportional to the sine of the zenith angle. The sine of 66.5° is 0.917, and the sine of 19.5° is 0.334. Thus, in December we have 2.747 times as much atmospheric attenuation as we do in June at high noon. On occasions when the sun does come out in December, you require no fancy instruments to tell you that the sunshine does not contain as much heat as it had the previous June.

Just as significant from both the weather standpoint and the solar electricity standpoint is the fact that during June the sun will be high enough above the horizon to provide useful energy for about 13 hours, whereas in December, even on a perfectly clear day, the sun will be far enough above the horizon to provide useful energy for only about 5 hours. In Oslo, Norway or Stockholm, Sweden, it would probably not be possible to capture a single hour of useful energy during the entire month of December. Conversely, at Lagos, Nigeria, a solar electrical system would probably provide useful energy about 11 hours per day all year round.

Each of these factors has a powerful effect upon the design of a useful solar electrical system. The 23.5° inclination of the equator means that a really efficient solar collecting system must be prepared to tilt in a north-south plane through an angle of 47° in order to accommodate training on the sun with change of season. For higher latitudes it must also work with strongly attenuated sunlight over short working hours. The expense is also raised by the fact that the weaker sunshine is accompanied by shorter working hours and more hours of darkness which require electrical illumination. These factors combine to multiply the storage requirement. In a practical, household solar electrical system—required to be completely self-sufficient in a northerly climate—the cost of the energy storage system would completely dominate the cost of the total system.

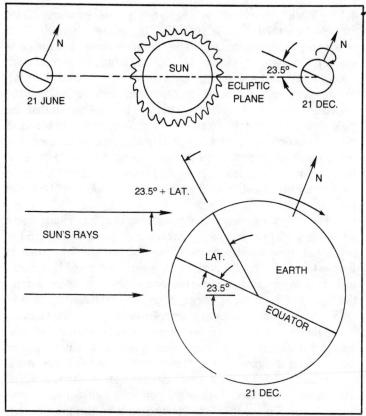

Fig. 10-1. Geometry of the sun-earth system.

JUST HOW MUCH ENERGY DOES THE SUN GIVE US?

The effective temperature of the sun as measured from the earth is 6150° K. The effective subtended angle of the diameter of the sun as viewed from the earth is a little more than a half-degree and the diameter of the earth viewed from the sun is 4.09×10^{-3} degrees. The earth intercepts about one two-billionths of the energy radiated by the sun. Of this fraction about half is radiated back into space by our atmosphere. We actually receive about 10^{18}kWh per annum. About 30 percent of this is reflected into space by plants during the daytime. The rest is distributed about as follows:

Night time radiation	$.7 \times 10^{17}$kWh
Evaporation of water	4.5×10^{17}kWh

On the average, plants utilize only about a thousandth of the solar energy incident upon them. Land vegetation consumes about

4.1 × 10^{13} kWh and marine vegetation consumes about 33 × 10^{13} kWh for a total of 3.7 × 10^{14} kWh. On the average, one kWh of solar energy can produce about 0.067 lbs of vegetation and the same amount of oxygen. It is estimated that the largest amount of solar energy that could be produced from a site in Arizona is 328,000 kWh per acre per year annum. If the same area were farmed and the vegetation burned to produce electricity, it is estimated that the yield would be about 44,700 kWh per acre per year. In terms of electrical energy production the direct solar process is more than seven times as efficient. The burning of the plants would just about consume the oxygen produced by the photosynthesis of the plants; the burning would just about liberate the same amount of CO_2 as had been absorbed.

The 328,000 kWh per acre per annum is an interesting figure since this works out to an average output of 37.42 kW. At 1.6 kW per home, this would support 23 homes. Even if the output in Rochester were only 10 percent as great, this would be enough electricity to support more than two homes per acre, which is about the density of the suburban areas. It has been estimated that if all of the wind in the world that was practical to employ were put in use, it would supply only about 3 percent of our current electrical demands. Similarly if all the water in the world practical for hydro power generation were employed, it might meet about half of the world energy demand. In the US it is estimated that about 40 percent of all the practical hydroelectric power is currently running, and it produces about 5 percent of our electricity. If all the practical hydropower in the US were put into service, the share for hydro would rise only to 12.5 percent.

The rest of the world is not better blessed in terms of potential hydropower; it is simply that the rest of the world uses so much less electricity per capita. The US is considered to have 13 percent of all of the potential hydropower in the world and of that fraction some 40 percent is up and running right now. As the standard of living rises through the remainder of the world, the shortfall of hydropower compared to world demand will probably increase.

Among the renewable resources, only in solar power do we find the potential for an energy source capable of supplying more energy than is used.

If we define the standard of living of people as being related to their energy consumption, then the US currently has about a 2.4 kW per capital standard of living. This is just about split equally three ways between electricity, transportation, and heating (for homes, industry, and all other purposes). Suppose that of the 4.5 ×

10^{17}kWh per annum that is used by the earth to evaporate water from the oceans we were to acquire just 0.1 percent or 4.5×10^{14}kWh per annum. Dividing by the hours in a year gives a continuous yield of 2.90×10^{10}kW. This would supply 2.4 kW to 12.1 billion people; which is more than three times the estimated world population. With a relatively negligible reduction of energy input into the oceans, we could more than provide a 2.4 kW standard of living for every man, woman, and child on the face of the earth!

However, before we become too ecstatic about this possibility, let us consider some of the ramifications. The area of the earth is about 6.6×10^7 square miles, and of this some 5.3×10^7 square miles is ocean. A tenth of one percent of the oceans is 5.3×10^4 square miles. If we were to assume that the system had an overall efficiency of 33 percent and that we had three floating power islands in the Atlantic, Pacific and Indian Oceans to provide round-the-clock power generation, then each island would have to be 230 statute miles on a side. This is larger than a great many of the countries on the earth! Just figuring a way to assemble, regulate, and distribute the power from these behemoths would require the work of generations of engineers. The construction would require a significant fraction of all of the iron and copper on earth, and there probably isn't that much wealth. In addition to this, the ecological impact of shadowing the waters in an area 230 miles on a side is at very least a subject for speculation. At the very least, the local climate could be expected to vary, and there are some who maintain that violent climatic changes like an ice age can be triggered by effects no larger.

Perhaps even worse is the political aspect of such a concentration. With the entire world dependent upon the supply of electricity of such an organization, the opportunity for tyranny would be unprecedented. Any nation could be instantly brought to a grinding halt by simply turning off its electricity—its sole source of energy. Within a matter of days, the lack of transportation, refrigeration, water pumping, heating, and lighting would bring starvation and chaos. The presence of such a disciplinary weapon would probably bring instant acquiescence to even the mention of a threat. Any attempt to construct such a worldwide power generation and distribution system would probably be met with a huge and violent opposition on the political front.

From the standpoint of the utility companies, the use of solar energy has a questionable practicality. In the perpetually sunny southern locations, the energy is available during only a portion of the day. In the northern climates, it may be essentially absent for weeks or months at a time. The possibility exists that electrical energy can be shipped over large distances. However, very large power transmission lines are expensive and difficult to operate. They also tend to be lossy. Current efforts to construct large intertie lines have been plagued by political opposition from environmental groups.

As a practical matter, electricity in commercial quantities is completely volatile. It must be consumed as it is made. With the exception of pumped water storage, there is currently no practical way of storing electrical energy. At the present writing, the facilities for pumped storage tend to run an overall efficiency of about 60 percent and have an installed kV-a rating of less than one percent of the national average requirement.

Because the entire hemisphere is dark at times, this boils down to the requirement that the complete generating facility would have to be duplicated. Since solar plants are estimated to run $12,000 (installed) per kV-a versus $12 to $120 for fossil or nuclear plants, this would mean a large multiplication of the capital investment to provide a facility that would be useful mainly for assuming the peak loads caused by air conditioners. Unless or until an economic mechanism for bulk power storage is perfected, the generation of large quantities of solar electrical power will be confined to very sunny locations. A similar analysis applies for home solar electricity, particularly in the northern climates. Solar power is present during only a fraction of the day in any location. In the northern climates, the fraction may fall to a quarter of the hours and the energy available may fall to something less than 10 percent on clear days and to zero on cloudy days. In Rochester, New York, it seems likely that the months of November through April would yield only about three to four percent of the electricity available from a given installation during the months of June through August. For such areas, it seems likely that a storage system capable of capturing energy in June and holding it until next April or May would be a requirement for a solar-only system. Of course, areas like Arizona, New Mexico, and Utah would require storage for only a matter of a half-day or day. Places like California and Florida, which have bright sunshine year round and a southerly location but are afflicted with rainy periods of a week or so, would require something on the order of a week of storage for a solar-only system.

SOLAR ENERGY RECOVERY

As noted toward the close of Chapter 3, the present monocrystaline solar cells are too prohibitively expensive to permit their use as a practical source of solar energy in the home. If some form of focusing or concentrating mechanism were applied, the yield per cell could be improved to some extent. However, the poor efficiency of the cells means that the majority of the energy incident upon the cell is wasted in heat. Because the cells themselves can easily be destroyed by overheating, the amount of improvement attainable in this direction is very limited. In weak sunshine or on a cloudy day, the focusing mechanism can be used to bring the output of the solar cell back to about the level that it would have had in bright sunshine. Yet in bright sunshine, any significant increase in output due to focusing would be bought at the cost of overheating and damage to the cell.

If the solar cell is too expensive, then how may we turn the energy contained in sunshine into electricity? Although a number of experiments are being pursued in which photochemical reactions may be used to "fix" solar energy into some form of fuel, at the present writing there are only two established and practical mechanisms for transforming solar energy into electrical energy:

- Using the sunshine to grow plants which may subsequently be used as fuel in a heat engine.
- Using the solar energy directly to operate a heat engine.

THE HEAT ENGINE

In order to understand some of the rationale behind the discussions to follow, it is necessary that we have some understanding of the properties of heat engines in general, since these have a great deal to do with the construction of a usable solar plant. Although Hero of Alexandria demonstrated a working reaction turbine in about 150 BC and Leonardo Da Vinci had a basic knowledge of the power available in steam, it remained for Newcomen to develop the atmospheric engine in 1712. In this engine, a vertical cylinder—open at the top—was fitted with a piston suspended by a chain from a teeter-totter arrangement with a counterweight heavy enough to lift the piston. Steam at atmospheric pressure was allowed to flow into the cylinder and the steam valve then closed. A jet of cold water was then squirted into the cylinder which cooled and condensed the steam. Atmospheric pressure then forced the piston down. These engines were slow, cumbersome, very inefficient, and relatively feeble. A huge engine with a four-foot diameter piston and a six-foot stroke operating at 10 strokes per minute would produce only about three hp.

James Watt realized that if the engine were made to operate by introducing the steam into the cylinder under some pressure greater than atmospheric and then allowed to expand, pushing out the piston, and finally allowed to condense and cool in another vessel, he could obtain a far more powerful and efficient engine. The early Watt engines reduced the steam consumption for a given amount of work by 75 percent compared to the Newcomen engines. The Watt engine was aided by the development of the machine boring mill by John Wilkinson, which made it possible to obtain much better fit between the piston and cylinder. By 1783, Watt had developed an engine capable of supplying rotary output. The steam engine very quickly became a commercial success since it was introduced at a time when virtually every available water power site in England was occupied and English industry had expanded to the point where cheaper and more abundant power was a virtual necessity.

It is noteworthy that our number for horsepower was developed by James Watt. Watt found that on an eight-hour shift, pumping water from a Welsh mine by walking on a treadmill, the horses averaged about 550 ft-lbs of work per second. The number was used to tell the potential buyer the number of treadmill horses the engine would replace.

The use of steam grew rapidly, and there were great inducements for inventors to improve the steam engine. By 1815, Olliver Evans of Philadelphia had built an engine operating at a pressure of 220 lbs per in^2. This engine was used to power a self-propelled dredging machine which walked on land and paddled itself through the water. The machine was called the Oruckter Amphibolos. This machine was one of the very first self-powered vehicles.

While the development of the steam engine leaped ahead under the hands of practical men like Evans, another group of men were studying the basic processes of heat, expansion, and the behavior of gasses, liquids and solids. It was the basic contributions to understanding of the processes that made the development of the efficient heat engine possible. Some of these principles must be understood if one wishes to have some understanding of the operation of a practical solar electrical plant.

CHARLES'S LAW

Figure 10-2 shows a very simple experiment that can be performed at home. A glass tube of uniform bore is heated in a gas flame and the bottom sealed. While the bottom is still hot, the top is dipped into some water containing a bit of food coloring. The contraction of the cooling air inside the tube will cause a drop of water to be sucked up. This water drop forms a freely moving piston or seal on the tube, and the pressure inside the tube will be equal to the atmospheric

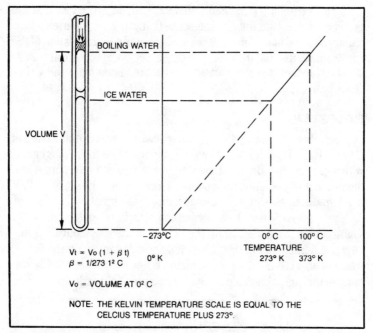

Fig. 10-2. Charles's law.

pressure outside of the tube. The volume of the gas inside the piston will be proportional to the height of the bubble.

If the tube is first dipped into icewater and the height of the bubble recorded, and then dipped into boiling water and a second height recorded, a graph like that on the right of the figure may be constructed. An interesting feature of this graph is the fact that the curve may be extrapolated to indicate that the volume would be zero at $-273°$ C. Over the range from $0°$ C to $100°$ C the graph will prove to be quite linear, and the coefficient B will be found to be 1/273 to reasonable tolerances. Stated differently, the gas will expand 1/273 of its volume when heated from $0°$ C to $1°$ C and will contract the same amount, if cooled.

CONSTANT VOLUME BEHAVIOR

The apparatus shown in Fig. 10-3 is a bit more sophisticated. In this device, the gas is contained in a reservoir sealed in by a mercury column. The pressure within the reservoir may be determined by measuring the height of the mercury column above the surface of the mercury in the mercury reservoir. If the mercury reservoir is much larger than the bore of the mercury tube, the volume of the gas is held constant to a first approximation.

159

The surprising thing about this experiment is the fact that the graph of the results tells us the same thing that the Charles's Law experiment did in a somewhat different way. An extrapolation of the results indicates that the pressure of the gas would be zero at $-273°$ C. The constant B is also identical to that obtained from the Charles's Law experiment.

BOYLE'S LAW

Another of the significant properties of gasses is enunciated in Boyle's Law. Figure 10-4 shows a cylinder and piston arrangement with a gas captive. Boyle's Law states that a constant temperature, the product of pressure and volume is a constant. The curve of P/V is a hyperbola as shown in the accompanying graph.

In Fig. 10-5 we show something a little bit different. If we slightly relax the pressure on the piston, it will move outward to a new position. In doing this it has done some work since the force, which is equal to the area of the piston times the pressure in lbs/in^2 moving through a distance ds represents physical work. In order to

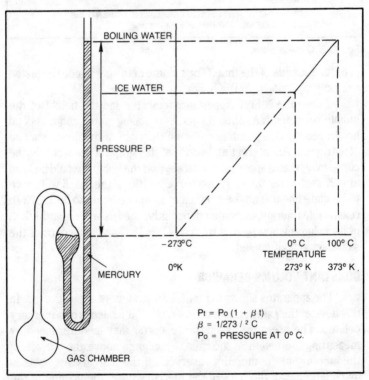

Fig. 10-3. Constant pressure.

160

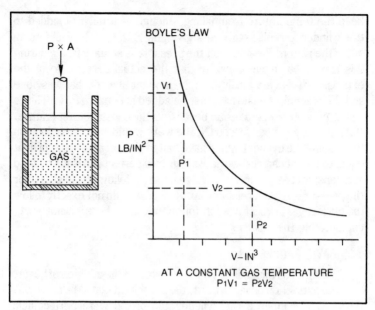

Fig. 10-4. Boyle's law.

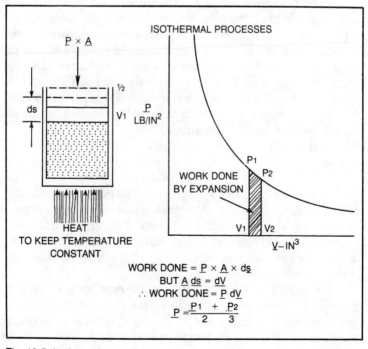

Fig. 10-5. Isothermal processes.

keep the temperature from falling, enough heat must be added to the cylinder to replace the work given up to the mechanical forcing out of the piston. We note that the pressure is actually falling during this travel, but if we consider that the actual percentage of the expansion dV is very small, then the amount of work that was done and the amount of heat that must be added is equal to PdV. This is called an *isothermal expansion* because of the constant temperature. If the piston had been forced IN, the work would have been done on the gas and the cylinder would have had to give off an amount of heat equal to the mechanical work. As long as a perfect gas is maintained at a constant temperature, it will reasonably follow Boyle's Law for the pressure-volume relationship. Real gasses do not exactly follow the perfect gas laws. However, for our purposes here it is not worth complicating the picture.

ADIABATIC PROCESSES

When a gas expands or contracts without the addition of heat to it or the extraction of heat from it, the process is said to be *adiabatic*. This would happen if the cylinder and piston were perfect heat insulators so that no heat could enter or escape during expansion or compression. In actual practice, certain processes which take place rapidly can approach this performance. For example, the air escaping from a balloon or tire which was at the ambient temperature will be perceptibly cold since it was expanded and has not had time to acquire heat from the atmosphere. Similarly, the barrel of a tire pump or the cylinder of an air compressor will become very hot when in use because the mechanical work involved in the pumping is partially converted to heat, which does not have time to radiate off during the brief compression stroke.

In an Adiabatic process, the equivalent of Boyle's Law is as follows:

$$P\,V^\alpha = \text{a constant}$$

The value of the exponent α is determined by the molecular structure of the gas. For most of the common gasses, such as hydrogen, oxygen, and nitrogen—which are diatomic—the value of α is 1.40. Some of the inert gasses, such as helium and argon, are monatomic and have a value for α of 1.66. Air which is composed mostly of diatomic gasses has a value of $\alpha = 1.40$.

Physically what this means is that in the rapid compression of a gas where the heat has little time to escape, the pressure winds up somewhat higher than it would have if the compression had taken place slowly enough for the temperature to remain constant. Thus an adiabatic compression requires more work than an isothermal

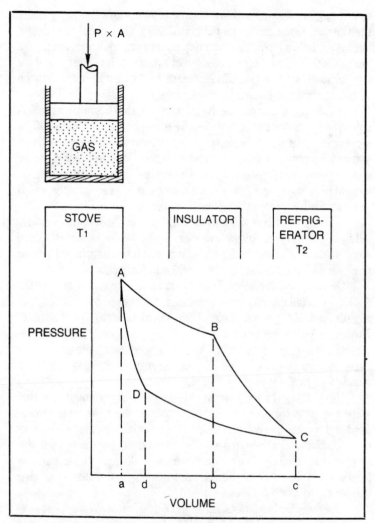

Fig. 10-6. The ideal heat engine—the Carnot cycle.

one. If the compressed gas is allowed to sit in storage for any period of time, it will come to equilibrium with the surroundings and the added heat will be lost.

Actual physical processes are never exactly isothermal or adiabatic but tend to lie somewhere between the two.

THE IDEAL HEAT ENGINE

In about 1830, a French physicist, Sadi Carnot, described an ideal heat engine. The efficiency of this engine cannot be exceeded by any engine working within the same temperature limits. Actually, the Carnot engine cannot be built. However, it is useful to consider because it tells us exactly how much energy we could extract from a heat engine if all of the processes were perfect. It is also useful for our purposes since it graphically illustrates the requirement for each of the major components of any heat engine.

Figure 10-6 shows the basic parts of the Carnot engine. A piston and cylinder that are perfect heat insulators are capped off by a cylinder head that is a perfect heat conductor. There are three stands on which the cylinder can be placed. The first is a stove or source of heat at temperature T_1. The second is an insulator which is incapable of transferring any heat. The third is a refrigerator which can extract heat from the cylinder at temperature T_2.

Starting with the cylinder standing upon the stove, the piston is held near the bottom of the cylinder. The cylinder contains a very small volume of gas at high pressure and at a temperature T_2 since the cylinder head transmits heat without restriction.

On the pressure versus volume diagram the unit is at position A. Upon reducing the pressure slightly, the piston moves out slightly and the gas expands. This would tend to cool the gas. However, heat flows unimpeded through the cylinder heat maintaining the gas at temperature T_2. The gas expands along the isothermal curve AB doing work by forcing out the piston. The gas absorbs a quantity of heat Q_1.

Next, the cylinder is moved to the insulating stand. On this stand the insulator prevents further heat transfer through the cylinder head and the gas undergoes a further expansion. This expansion is adiabatic, and the expansion is allowed to continue until the temperature falls to T_2. On the P versus V diagram the expansion follows the curve BC. During this expansion the piston is also delivering work to the outside world.

The cylinder is then placed on the refrigerator. Work must now be delivered from the outside world to force the piston in. The heat Q_2 generated, flows freely through the cylinder head to the refrigerator; the compression is isothermal along path CD. This compression requires minimum work, and the gas remains at temperature T_2.

At point D, the cylinder is again placed on the insulating stand, and the piston absorbs work from the outside world to adiabatically compress the gas back to the volume and pressure of point A. The work delivered to the outside world is proportional to the area circumscribed by ABCca, whereas the work required to get the gas ready for the next cycle is proportional to the area circumscribed by

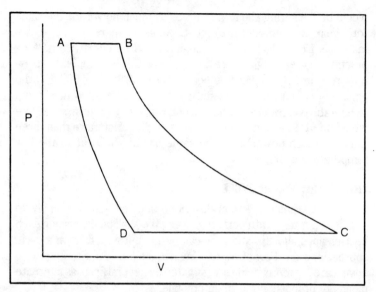

Fig. 10-7. The vapor cycle. The area under the curve represents the work done by the engine.

CDAac. It can readily be seen that the net mechanical work delivered by the engine is proportional to area ABCD. The engine has delivered net work from the process and has transported heat from the stove to the refrigerator in the process. The Carnot Efficiency is given by the formula:

$$\text{Carnot Efficiency} = \frac{T_1 - T_2}{T_1} = \frac{Q_1 - Q_2}{Q_1}$$

where the temperatures are measured from absolute zero or $-273°$ C (the temperature where the pressure and volume of a perfect gas would be zero).

The significance of this relationship is very great. Suppose that we operate the Carnot engine with T_2 at the boiling point of water (100° C) and the refrigerator at a room temperature of 25° C. The Carnot Efficiency would be:

$$\text{Carnot Efficiency} = \frac{(100 + 273° \text{ C}) - (25 + 273° \text{ C})}{(100 + 273° \text{ C})}$$
$$= 0.201$$

At the very most, using the ideal engine we could recover only about 20 percent of the heat energy.

By comparison let us suppose that an ideal converter would permit us to operate the stove at the temperature of the sun or 6150° K while the refrigerator remained at room temperature. The Carnot Efficiency in this case would be 95.2 percent. The engine

165

would be delivering about 4.74 times as much power for the same total heat (not temperature) input. Actually, there are no known materials from which such an engine could be constructed. As a practical matter, engines are limited to operating temperatures on the order of 500° C which is equivalent to 773° K or 932°F if the engine is continuously exposed to the hot gasses. With a room temperature refrigerator the Carnot Efficiency of such an engine would be 61.4 percent. This engine would extract more than three times as much power from the same quantity of heat as the low temperature example.

THE CARNOT VAPOR CYCLE

The Carnot cycle thus described and the gas laws apply to engines operating with perfect gasses. We shall be discussing such heat engines shortly, and they can be made to operate on a cycle approaching the Carnot cycle. Engines such as the Sterling cycle engine are currently being investigated by several firms as alternate means of propulsion for an automobile.

By far the more common is the conventional steam or vapor cycle engine. Because of the differences in behavior the P versus V diagram of such an engine is considerably modified.

Consider an engine identical to the Carnot engine shown earlier, except that the cylinder is filled with a liquid at the boiling point at point A. Figure 10-7 shows that the addition of heat from the stove causes boiling under constant temperature and pressure conditions. At point B, the vaporization of the fluid is complete and the cylinder is transferred to the insulating stand. The vapor then adiabatically expands from B to C, while the temperature drops to T_2.

The cylinder is then moved to the refrigerator and the volume is reduced at constant temperature and pressure. At this point some of the heat of vaporization which was acquired to boil the liquid going from A to B is rejected to the refrigerator. This corresponds to path CD. The vapor is restored to liquid form. The cylinder is then moved to the insulator and the remaining small compression restores the condition to point A.

Both this vapor cycle and the gas cycle previously described are reversible. The vapor cycle actually closely approximates the performance of a real steam engine or household refrigeration unit. Because of the high heat of vaporization of the working fluid compared to a gas, the volume to be handled can be smaller in the vapor cycle engine than in the gas cycle engine.

SUMMARY

The nature of heat engines and the properties of gasses and vapors make it important to operate the engine at as high a temperature as is practical for the materials and techniques employed in the machine construction. For this reason, the construction of the solar electrical plant differs widely from the construction of a solar heating apparatus. In the latter, passive solar panels are frequently operated with a temperature rise of only 8° to 10°F. For this small rise an efficiency of only 3.2 percent would result, and the plant would have to be unacceptably large to produce any electricity. Compared to a high-temperature system, the power output would be only one-twentieth as great.

These factors add up to the fact that a practical solar electrical system must employ concentrators capable of producing the high temperatures required for reasonably efficient operation. The next chapter is devoted to a discussion of such concentrators and mechanisms for tracking the sun.

Solar Collectors

11

The discussion of the previous chapter was intended to show something of the geometry of the earth-sun situation and to emphasize the requirement for the development of the highest possible temperatures in the solar collector apparatus if the object is to operate a heat engine for the generation of electricity. In this chapter we shall examine the fundamentals of the design, construction, and testing of solar collectors suitable for the operation of heat engines. In order to understand the principles involved, we shall have to treat a few selected topics in the areas of optics, geometry, and electronic servo control. We shall attempt to go no further into these topics than is absolutely necessary to provide an understanding of the factors involved in the design of a workable solar collector.

OPTICS

A solar collector has certain fundamental similarities to an optical or a radio telescope. One of the prime goals is the collection of the largest possible amount of power which is incident upon the *aperture* (or collecting area) of the instrument and directing it to a specific point or area where it can be used in some manner. It should also be equipped to track the body under consideration with minimum complexity so that the image is stationary with respect to the energy collecting medium.

There are also certain rather fundamental differences. Both the optical and the radio telescope are designed to capture the largest possible amount of the energy incident upon the instrument aperture and to provide the highest possible *resolution* (the ability to distinguish between two small adjacent bodies) so that closely spaced bodies can be separated. Both are measuring engines, rather than power collecting engines.

In order to provide the desired resolution, both the radio and the optical telescope must have *path length errors* (distance from object to reflector to collecting apparatus) which are small with respect to the wavelength. In the optical telescope, the wavelengths of visible light range from 0.072 to 0.038 × 10^{-6} meters. For the radio telescope the useable range of wavelengths is from about 1.5 × 10^{-2} meters to something on the order of 150 meters. The wavelengths on either side and the gap are effectively shut out from an energy standpoint by the nature of our atmosphere. From a practical standpoint, the energy available from moonlight and starlight are insignificant for the production of home electricity. Only the energy from the sun is of any practical significance. Again, as a practical matter, it is only the energy lying between 3 and 0.3 × 10^{-6} meters that has any significant contribution to the generation of household electricity. The shorter wavelengths are effectively absorbed by our atmosphere, and the longer wavelengths are too feeble to contribute any meaningful amount of heat to a heat engine.

Another factor which contributes to the differences in design of solar collectors compared to telescopes stems from the fact that, unlike stars which have no measurable subtended angle, the sun has an appreciable subtended angle. The diameter of the sun measured from the earth subtends an angle of about a half a degree in the visible wavelengths and about 0.66 degrees at the longer wavelength limit. The reason for the variability is the fact that while the solar corona does not emit much visible light, it is still a copious emitter of longer infrared wavelengths and radio wavelengths.

In a solar collector, it is not a particularly worthwhile objective to obtain a well defined image of the sun, but rather to simply collect the largest reasonable amount of the energy. For this reason, it is not necessary to build the solar collector with the extreme precision of optics which is required for an optical telescope. As will be noted shortly, advantage is usually taken of this fact in the design of solar collectors because of the economies involved.

There are three basic types of optical telescopes. The *refracting, reflecting,* and *catadioptric*. The refractor consists of a lens type *objective* (the light-gathering lens at the front of the telescope) and an eyepiece for viewing the image. The reflecting type makes use of a mirror for the objective. The catadioptric type employs a combination of lens and mirror for the objective to reduce spherical aberration and to reduce the tube length for a given magnification. All three types are used for solar collectors, although the catadioptric is generally employed only when available as surplus because of the expense. We shall begin by discussing the refracting type first since this is the easiest type to understand and visualize. Before we do this, it is necessary to discuss a few topics which are necessary for the understanding of some of the behavior of the optical components.

ANGLE OF INCIDENCE = ANGLE OF REFLECTION

Figure 11-1 illustrates the reflection law. It assumes that:
- Light travels at a fixed velocity in free space and at essentially the same velocity in air; noted on the diagram as C_o.
- The propagation of light is in a direction normal to the wavefronts.

The illustration shows a train of light waves travelling from left to right across the picture. The velocity of light in the direction of travel can be resolved into components tangent to the surface and perpendicular to the surface. These are noted as V_t and V_p, respectively. If the light waves collide elastically with the surface, we see that V_t will be unchanged, but V_p will be reversed in sign. Thus the little triangles which constitute the resolution of C_o will be mirror images of one another. Therefore the angles of incidence and reflection must be identical!!!

Because of the immense difficulties involved in making such a lens of optical quality, this type of lens has never been popular for astronomical or photographic work. However, the Fresnel lens has found wide popularity for lighthouses, searchlights, railroad lamps (and lately for magnification of TV images) where a very large, short focal length lens of not-too-great image clarity is required.

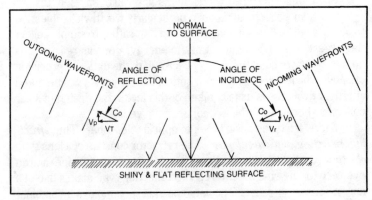

Fig. 11-1. The reflection law.

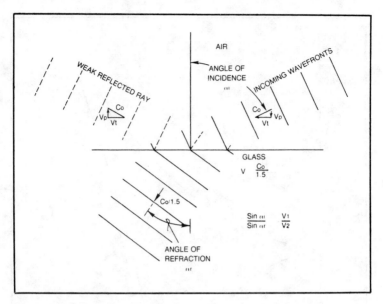

Fig. 11-2. Snell's law.

It is obvious that a certain fraction of the light in the cracks between sections will be lost. However, light gathering efficiencies in excess of 80 percent are readily achievable. Fresnel lenses molded by a plastic replication process have found wide popularity as solar collectors. These lenses can easily be molded with aspheric correction to provide very large light gathering power and very small focal length to diameter ratios for a hot, intense image.

The lens has an advantage over the typical reflector in that the lens need not be held with any great precision if image quality is not a prerequisite. The lens-type solar collector can therefore be built lighter and perhaps cheaper than a reflector type of equivalent size.

THE REFLECTOR TYPE SOLAR COLLECTOR

Figure 11-6 shows a section of a spherical surface with parallel rays of starlight incident upon it. For the rays striking the surface close to the axis it may be seen that the angle to a point halfway between the vertex V and the center of curvature C make an angle very close to the angle between the radius of curvature and the incoming ray. In fact, the approximation is so good that it will yield astronomical quality images if the diameter of the reflector portion used is not greater than one-tenth of the focal length of one-twentieth of the radius. However, in the outer portions of the sphere, the angles with the subscript "3" show that the quality of the

171

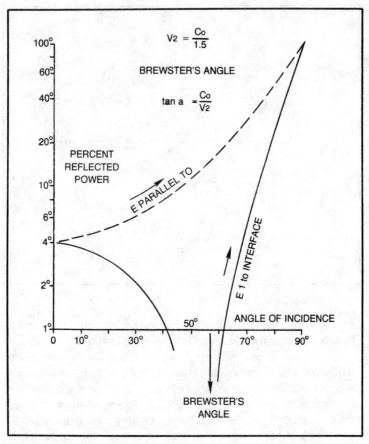

Fig. 11-3. Reflections from an interface.

focusing breaks down rapidly. This is termed *spherical aberration*. As with the thin lens, the long-focal-length reflector is of little value for solar collecting. Yet there is a relatively simple aspheric correction which can be made to the surface to eliminate this problem. Refer to the figure to see that the problem with the focusing from the outer portions of the reflector is due to the fact that the reflector curves around too fast. A parabola, which shows an increasing radius of curvature with distance from the vertex, can be shown to perfectly focus both the inner and the other rays for a point source. In astronomical work, other factors such as image distortion tend to limit the useable designs to a mirror diameter one-sixth or one-eighth of the focal length. But with radio telescopes and radar antennas, the use of diameter-to-focal-length ratios as high as four is common. Figure 11-7 shows such a parabolic reflector.

It is noteworthy in this design that the heat collecting apparatus must be able to accept the focused energy over a wide range of angles. If the diameter of a reflector of this same focal length were carried to 12 units, the rays from the periphery of the reflector would be reaching the focus at right angles to the central ray. The parabola shown is still a very "fast" objective. To the photographer it would have an "f-number" of 3/10.

For photographic or astronomical telescope purposes, such a very "fast" reflector would be of little use. The distance from the point of reflection to the focus for a point on the periphery (Y=5, X=2.08) is 5.08 units compared to the distance for a ray along the x axis of 3 units. This means that the image of a source of finite extent, such as the sun, would be larger by the ratio of 5.08/3 for the rays from the periphery compared to central rays. This would hopelessly blur the image. However, for applications such as a radar reflector or a solar collector where no attempt is made to achieve a well defined image of the source, this does not represent a serious fault. The very short focal length provides a compact mechanical structure which can be made very rigid.

Workable solar collectors can usually be made "faster" in the reflecting form than in the refracting form. This is due principally to

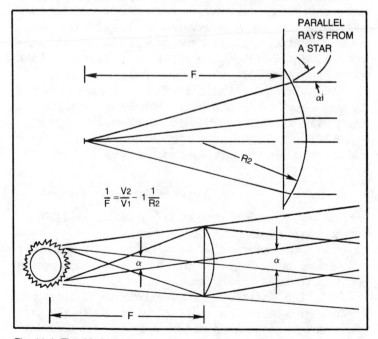

Fig. 11-4. The thin lens.

the angle-doubling properties of the incidence/reflection law. Even in the relatively extreme design shown, the angle of incidence for rays parallel to the axis is only 50.19°. For a refractor with the same diameter and focal length, the rays at the periphery would have to have been bent through an angle of 59°, which is considerably more of a problem.

SNELL'S LAW—REFRACTION

The illustration of Figure 11-2 shows a ray of light or a parallel wavetrain incident upon a piece of glass. The formula at the bottom was developed by a Dutch astronomer and mathematician, Willebrod Snell, in 1621 to describe the action and is usually referred to as Snell's Law. It states simply that the sines of the angles of incidence and refraction are proportional to the ratios of the velocities of propagation in the two media.

If the wavefronts are pictured as a number of ranks of soldiers and the glass/air interface is pictured as a junction between a muddy field and smooth pavement, the analogy explains the action. As the soldiers at one end of the rank encounter the mud they are slowed and begin to fall behind the rest of the rank. The rank acquires a kink in the middle. When the soldiers all have finally entered the mud, the rank is again straight. However, the column has been bent to a new angle of travel and is now proceeding at a slower speed. The soldiers are also spread farther apart in the rank. Snell's Law is useful in the fact that it describes the operation of lenses and prisms and also shows some of the limitations on the design of these items.

One factor which is of interest is the fact that since the sines of the angles can have only the values between zero and one (the negative values have no physical significance here) there is an angle beyond which the light cannot pass from the denser to the less dense medium but is instead totally reflected. This angle is given by the formula:

$$\text{Sin } \alpha_i = \frac{V_2}{V_1} \quad \textbf{Angle of Extinction Formula}$$

For glass with an index of refraction of 1.5, this angle is 41.8°. No transmission is possible, at any larger angles of incidence. This is the effect which makes a 45-45-90° prism behave as if a mirror existed along the hypotenuse when the prism is viewed through one of the shorter faces.

In Fig. 11-2, you will note that a weak reflection was noted also. The reflection follows the angle of incidence/reflection law. Whenever light encounters a medium of different index of refraction, a certain percentage of the light is reflected. Sunlight is randomly polarized and can be thought of as having two components with polarization perpendicular to and parallel to the interface between the media. The amount of reflection is different for the two components and is a function of the ratio between the indices of refraction and the angle of incidence as well. The curve of Figure 11-3 shows the variation of the reflection coefficients for the two components. It may be seen that the component polarized normal to the interface starts at normal incidence with a reflection coefficient of about four percent and actually drops to zero at an angle called the Brewster angle. It then rapidly climbs to unity. This property was discovered in 1816 by Sir David Brewster. At this angle 100 percent of the reflected light is polarized. All of the normally polarized component is transmitted without loss and only the parallel polarized component is reflected. Beyond about 80°, the majority of the incident light is reflected and very little transmitted by the glass. These factors place certain limits upon the design of lenses and certain types of reflectors or mirrors.

THE SIMPLE LENS

Figure 11-4 shows some of the components of a very simple plano-convex lens. If the assumption is made that the lens is very thin, then the focal length of the lens is given by the formula shown. For a lens with an index of refraction of 1.5, the rays of light from a distant object will be brought to a focus at a point equal to about twice the radius of curvature of the lens. As the parallel rays of starlight strike the curved front surface of the lens at increasing radius from the center, they do so at an increasing angle of incidence, thereby causing more refraction. The rays through this mechanism all come to a focus or point of concentration P.

Unfortunately, the thin lens is of little use for solar collection. As shown in the lower portion of the illustration, if the original object has some significant angular extent, the image is spread over about the same angle behind the lens which makes it large and diffuse rather than compact and hot. The very assumption of a thin lens implies that the front spherical surface never makes a very large angle with the incoming parallel rays, so the focal length is large with respect to the lens diameter.

As the lens starts to become thick, the compensation between the angle of incidence and the point at which the rays strike the lens begins to break down and the lens no longer will function well to focus the rays. Under these circumstances it becomes necessary to make the lens aspheric or to make it curve less rapidly than a sphere does in the outer portions. Such lenses are actually made for use as condensing lenses in slide projectors and enlargers, but one that was large enough for solar collector use would be intolerably heavy. This problem can be solved by modifying the lens.

Figure 11-5 shows a modification technique. Suppose that you took a large glass hemisphere and cut the center of it out with a cooky cutter shaped tool and then proceeded to remove progressively larger annular rings. If the center plug and the rings were then stacked on a flat surface with round surface down, then glued together, and the projecting back ground off, you would obtain a figure which would appear in cross section like the illustration. The majority of the weight would be removed. However, the device would still have the properties of a thick lens, including the short focal length for maximum image concentration.

FRONT AND REAR SURFACE MIRRORS

In the discussion up to this point, it has been assumed that the reflection took place from the front surface of the mirror. Astronomical telescopes are usually constructed in this way to avoid the confusion of image shown in Fig. 11-8. The incoming ray will reflect some portion of the light from the first surface, refract into the mirror, reflect from the silvered (or aluminized) surface and refract back out of the glass with some of the energy reflected back in. If the angle of the returning ray exceeds the angle of extinction, the ray may be trapped within the glass. At the very least, the rear surface will give rise to multiple images. By itself, this is probably not too serious in a solar collector. However, the trapped rays will tend to concentrate toward the center of the mirror and contribute to heating of the glass. On a large solar collector, this heating could be sufficient to fracture the glass.

There is a distinct advantage to placing the reflecting surface on the rear side of the glass. This is the way that hand mirrors and rear-view mirrors are constructed because it protects the silvered surface from scratching. If the mirror gets covered with dirt or dust, it can be simply washed just as one would wash a plate glass window. By comparison, a front surface mirror on glass requires great care in cleaning to avoid scratching or even wearing off of the reflecting surface. In addition, the rear surface mirror has the advantage that the reflecting surface is protected from exposure to the atmosphere by the glass and is thus relatively immune to oxidation or tarnishing.

There are certain coatings which have been developed to protect the front-surface mirrors used in astronomical telescopes. These coatings are vacuum deposited to a thickness of a few molecules and because of their extreme thinness do not disrupt the

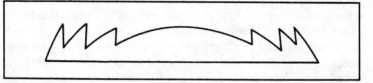

Fig. 11-5. The Fresnel lens.

image formation. However, they would be extremely expensive to apply to a reflector as large as a solar collector.

After a discussion of the requirements of a solar collector, we shall return to the relative discussion of first and second surface mirrors. At the present writing the lack of an inexpensive, durable, thin coating for large reflectors essentially precludes the use of first surface mirrors for practical solar energy collection. The surface would be too expensive to produce and too difficult to keep free from dust, dirt and tarnish (or oxidation and weathering).

THE REQUIRED SOLAR IMAGE

With either a lens or a reflector system, even the finest optical element would reduce the image of the sun only to a disc at the focus with a subtended angle measuring something on the order of a half degree measured from either the center of the lens or the vertex of

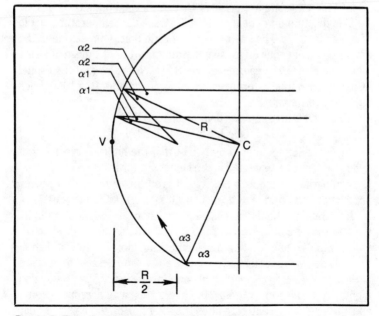

Fig. 11-6. The spherical mirror.

the parabola. If this were done, the image would attempt to assume a temperature equal to that of the sun or 6150° K, which is equivalent to 10,610°F. There is no known material which will remain solid at this temperature (or even liquid, for that matter). It is therefore necessary that the device receiving the image must be cooled by some mechanism since the temperatures are high enough to cause firebrick to dribble and run like water. In actuality, it is not necessary in a solar collection system to have the image collected down to a focused image; it is only necessary only to have enough concentration to obtain the desired heating of the working fluid for the heat engine.

Note: The shift over into English units is made here because the overwhelming portion of the engineering data available for boiler and steam engine design in the US is given in English units. The conversion factor for Kelvin temperatures to Farenheit temperatures is:

$$[(°K - 273) \times 9/5] + 32 = °F$$

On the absolute temperature scale in English units the temperatures are given in degrees Rankine. This is 459.4° higher than the Farenheit temperature. Absolute zero degrees Rankine is equal to −459.4° F.

$$°R = °F + 459.4$$

Therefore: $$[(°K-273) \times 9/5] - 491.4 = °R$$

Most heat data are given in British thermal units (BTU). One BTU is the amount of heat required to raise the temperature of 1 lb of water by 1°F. The calorie is the metric unit. One calorie is the amount of heat required to raise the temperature of 1 gram of water 1° C. There are 252 calories in one BTU. Note also that the capital "Cal." used in food references is a kilogram calorie. It is 1000 times as large as the cal.

One horsepower hour = 2546 BTU

One kWh = 3413 BTU

It has been estimated by NASA that the highest direct flux of solar energy to be expected in the United States is 330 BTU/ft^2/ hr. By comparison a modern coal or oil-fired boiler for steam power production will operate at a flux of 40,000 to 50,000 BTU per ft^2 per hr. If we take the boiler limit as being a reasonable engineering limit, we see that a concentration ratio of 150 would be about what we would want to use for this maximum solar flux. If the installation were to be used in a northern climate and it was desired to operate with a feebler solar input at maximum efficiency, we could use higher concentration ratios. However, this would require some aperture reducing control to keep the solar flux on the boiler or heat gatherer within safe limits during the summer months.

SIZE OF COLLECTOR

It is instructive to consider the size of reflector that we might require in order to produce a reasonable amount of home electricity. For our model system let us suppose that we wish to have the system operate at a solar flux level of 165 BTU per ft² per hr or one half of the maximum expected. For a 1-kW output with an overall efficiency of 0.06, we would have to use a collector with an area of:

$$\frac{3413 \text{ BTU/ kWh}}{0.06} = 165 \text{ BTU/ft}^2/\text{hr} \times \text{Area (ft}^2)$$

$$\text{Area} = 344.75 \text{ ft}^2$$

If the collector were round it would have a diameter of 21 feet. For our 3 hp equivalent, we would require 771.5 ft² and a diameter 31.3 feet.

The 3-hp model is selected because of the fact that even under the conditions of the equator, the solar plant will operate effectively for only a portion of the day and the need therefore exists to produce energy over and above the immediate usage for storage. If we only required that the unit produce the 2.24 kW (3 hp) at the peak solar energy level, the size of the collector could be shrunk to a fifteen foot diameter. This is a much easier and less expensive size. However, it would only here and there and now and then meet the requirements for home electricity—even on the sunny Arizona desert.

The design and construction of a 32-foot paraboloid of revolution presents some problems, but reflectors of this size are commonly constructed for radar, satellite terminals, and radio telescopes. These units are also mounted on tracking pedestals or mounts. The principal obstacles are therefore no insuperable engineering problems but rather the economic constraints.

The collection efficiency figure employed in the calculations may seem excessively pessimistic at first glance. However, it will be noted later that it is not. The figure was derived as follows:

Engine thermal efficiency	0.089
Engine mechanical efficiency	0.71
Boiler efficiency	0.974
Collector efficiency	0.974

The overall product of these is the 0.06 number used in the calculations. The engine mechanical efficiency is intended to include the friction losses, departure from ideal cycle, inefficiency in the generator, and the work done on the auxiliary pumping and tracking machinery. In a small plant, such as the unit under discussion, an efficiency of this order is quite high.

The boiler efficiency number is intended to include the heat losses between the boiler and the remaining steam system. The collector efficiency number is intended to include the amount of sunlight lost in heat in the collector itself and the amount which is not absorbed by the boiler, as well as a certain amount of scattering due to dirt or dust on the collector. Here again the efficiency cited is quite high.

It is obvious that an increase in these numbers would make it possible to operate the system with a smaller collector. Yet if all of the inefficiencies beyond the basic thermal efficiency were made equal to unity, the collector area would only be reduced by a factor of 1.48 and the collector diameter would be reduced by a factor of 1.22. The cost of collectors tends to run at something like the 1.3 power of the area. Therefore, the cost would fall by a factor of about 1.7 if the same quality of construction were able to provide the requisite improvement.

The purpose of this discussion is, of course, to illustrate the fact that it is probably economically warranted to go to considerable pains to obtain a relatively high efficiency in each of these departments. As with all engineering efforts, there must certainly be some point of diminishing economic return where the incremental improvement in performance is not balanced by the increased investment required. The history of solar steam installations is so limited that no rule-of-thumb guidelines are available, and the investigator is pretty much on his own to develop an economically feasible model.

THE HEMISPHERICAL BOILER

Both of the collectors discussed thus far have been figures of revolution about the focal axis. If the boiler is to be operated at the 50,000 BTU per ft^2 per hr rate, it is apparent that the energy incident upon the collector cannot be focused down into too tight a spot. It can be shown that a hemispherical target concentric with the focus of the reflector will have a nearly uniform heat input over the area illuminated.

For a reflector such as the one illustrated in Fig. 11-7 with a diameter of 10 units and a focal length of 3 units, the subtended angle of the reflector seen from the focus is a cone with a half angle of 79.6°. Thus the boiler should have the shape of a portion of a sphere with a subtended angle somewhat larger, perhaps 85°. The illustration of Fig. 11-9 shows a suggested arrangement in cross section. The water enters along the axis of the hemisphere and is constrained to flow between the dome shaped surface receiving the solar energy and the backing plate. The water flows to the circular

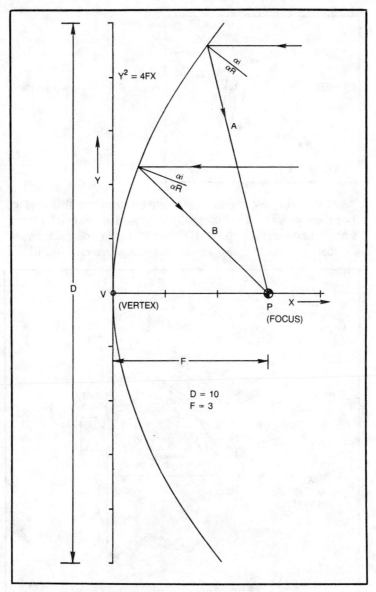

Fig. 11-7. The parabolic reflector.

manifold and then out the outlet pipe. If the front of the dome is relatively thin, there will be a relatively small temperature drop between the absorbing surface and the water surface.

An example here would be instructive. Let us consider a solar collector of 2.24 kW electric output rating operating in an incident

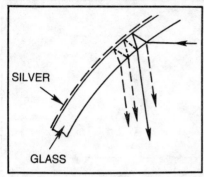

Fig. 11-8. The rear-surface mirror.

flux of 165 BTU per ft² per hr. For conservatism, we shall neglect the reflector losses. The boiler will be exposed to 130,600 BTU per hour. For our nominal 50,000 BTU per ft² per hr boiler design we see that we would require a boiler collecting area of 2.612 ft².

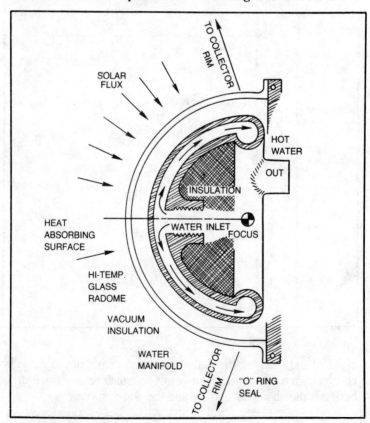

Fig. 11-9. The hemispherical boiler.

The formula for the area of a segment of a sphere cut out by a cone is:

$$A = 2\pi r^2 (1 - \cos \theta)$$

For our reflector, $\theta = 79.6°$. Using this formula, we find the radius of the boiler to be 0.71 feet or 8.55 inches.

The design of boilers is generally done under ASME design rules which have two factors of particular interest to us here.

- A boiler should be hydrostatically tested at 150 percent of the maximum rated pressure, struck a number of blows with a calibrated hammer and then hydrostatically tested to 200 percent of rated pressure.
- Except in special cases, the maximum tensile strength of steel is taken as 50,000 lbs per in^2.

Now, let us suppose that we wish to operate our boiler at 200 pounds per square inch gauge (psig). Because of the vacuum drawn on the outside of the boiler for insulation purposes, we would have a pressure differential of 215 pounds per square inch absolute (psia). To allow a bit of leeway, let us design the boiler for 450 psia. The projected area of a hemisphere of 0.71-foot radius is 1.594 ft^2, and the circumference is 4.475 feet. Multiplying the area by the pressure, we find that the boiler dome is holding a load of 49,341 lbs! Dividing this by the periphery of the boiler, the force works out to 918.8 lbs/per lineal inch. For a stress of 50,000 lbs per in^2, we find that the boiler dome could be made of steel 0.018 inches thick (about 1/64ths of an inch), or #24 B&S gauge.

The conductivity of steel is about 26 BTU per hr per °F per ft thickness. At our 50,000 BTU per ft^2 per hr input rate, the temperature rise across the heated portions of the boiler works out to 2.95°F. There would of course be a gradient brought about by the amount of temperature rise the water experienced in the flow from the center to the periphery of the boiler. This will be discussed in more detail later. However, it is worthy of note here that it is inadvisable to let the temperature gradient rise too high since this will cause warping and expansion stresses upon the boiler as well.

The message to be conveyed by the preceeding exercise is the fact that there are tremendous forces involved in the solar boiler. The design of such a unit should not be undertaken by anyone not thoroughly versed in mechanical engineering practices. The American Society of Mechanical Engineers Boiler Code should be followed wherever local codes and ordinances do not override it. Unless properly designed, installed, and tested, boilers are *dangerous*. Boiler explosions used to regularly take a toll in human life and property!

INSULATION

Of course, the rear surface of the boiler can be insulated from heat loss to the atmosphere by any of the common high temperature insulating materials such as fiberglass, asbestos, etc. However, the dome itself would lose a great deal of the hard won heat to the atmosphere if it were exposed. Unlike the back side, the dome must have free access to the solar flux being concentrated by the collector. This is probably most easily accomplished with a glass or pyrex transparent radome which is evacuated to minimize convection losses to the atmosphere. At the temperatures that the boiler will operate, the losses will be almost entirely due to conduction—and in an uncovered boiler—due to convection currents and wind currents in the atmosphere. Evacuation of the radome will reduce these by more than an order of magnitude.

TOLERANCES

One of the major factors affecting the practicality of a reflector-type solar collector is the question of the tolerances permissible in the construction. The reflector sizes thus far discussed are noticeably larger than the very largest optical telescopes. In order to maintain the image quality of these telescopes, it is necessary to maintain the "figure" of the mirror within a fraction of a wavelength of the design surface (which is not always a parabola). For this reason, one finds that the *cost* of these large telescopes, as well as the purpose, is astronomical.

By comparison, the solar collector does not have the requirement that the image be well defined. It is only necessary that the overwhelming majority of the energy strike the boiler. This point is important since even a fraction of the energy collected, if focused upon one of the boiler supports would rapidly overheat that portion of the structure. If the situation were severe enough, it could even melt the structure.

To examine this requirement, let us refer to Fig. 11-10. This illustration shows an ideal theoretical paraboloid with a flat mirror tile placed tangent to the paraboloid at the center of the mirror surface. The reflection of the sun from this mirror surface would spread about a quarter of a degree from the centerline. If we assume a paraboloid of 32-foot diameter with a focal length of 9.60 feet and a boiler radius of 0.44 feet, we find that the image of the sun would spread about 0.91 feet on each side of the centerline. If the mirror were perfectly aimed so that the center of the image of the sun lies on the radius connecting the boiler center (the focus of the paraboloid) and the center of the mirror tile, we would find that it

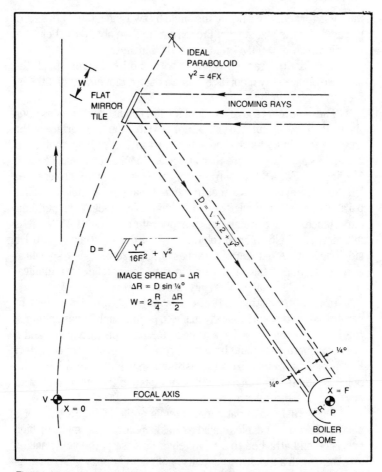

Fig. 11-10. An ideal theoretical paraboloid with a flat mirror tile placed tangent to the paraboloid at the center of the mirror surface.

was possible for the mirror to have a diameter of 0.258 feet and still keep the entire reflected image of the sun within a 30° cone on the boiler dome.

Next let us suppose that we allow the mirror tile to tilt to the extent that the image of the sun is just tangent to the sphere described by the boiler. For the tile at the $Y = 16$-foot point, we would find that the ray from the mirror had deviated through an angle of:

$$\frac{0.44 - 0.91 - 0.258/2}{20.83} \quad = \quad \text{arcsin } 0 = 0.61°$$

From the law of incidence and reflection we know that the angle through which the ray deviates is twice the angle through which the

185

mirror is turned. If we divide the angle by two and multiply the sin of the angle by the diameter of the mirror tile, we obtain the height of the tilt which is 0.00137 feet or 0.0165 inches.

If we try the same procedure with another mirror of half the size, we obtain a ray tilt of 0.783° and a mirror movement of 0.00088 feet or 0.0106 inches.

These numbers give us a feel for the required tolerances upon the reflector. The surface should not deviate from the surface of the true paraboloid by more than about 1/64 inch over a span of about 3.5 inches and by no more than about 0.01 inches in a span of about 1-¾ inches. For a 32-foot structure, these are very tight tolerances indeed. However, I can attest to the fact that it is possible to build a parabolic reflector of this size, with these tolerances, from personal experience. For a radar antenna operating at 6000 MHz, it is necessary to hold to such tolerances in order to attain a −30 Db sidelobe level. A number of antennas of this size with such specifications have been built in production runs for satellite communications, and radar usage. They are not very cheap.

It is noteworthy that the deviations apply to a best-fit paraboloid. Provision is usually made to permit slight adjustments of the feed location on radar antennas for best performance, and a similar adjustment would be a wise provision for a solar collector. For a collector of this size an adjustment span for the location of the boiler of about $66 1 inch should be adequate if the unit is built with anything like the tolerances above.

The actual use of flat mirror tiles is not at all impractical for a solar collector. The tiles could be laid upon a male mold of high precision and attached to a relatively imprecise paraboloid backing structure with a thixotropic (sticky, not runny) epoxy glue, such as car body compound used for repairing rust holes and dents. This technique of using a relatively imprecise backing frame and attaching it to a reflecting surface which is held to shape by a precise mold is frequently used in antenna fabrication. Since the mold can be used to make a great number of reflectors, it is economically advantageous to spend a substantial sum on the construction of the mold since this forms the surface contour for the reflectors to be produced. The back frames can then be made relatively imprecise at a considerable saving in construction cost. Obviously, this philosophy applies only when there is some volume production.

The use of the flat mirrors has another advantage in the fact that many of the objections to second surface mirrors are reduced or minimized if the mirror is flat and is not being used to focus a precise and well defined image. There is also the obvious economic advan-

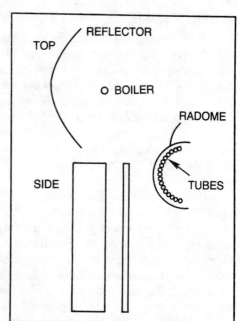

Fig. 11-11. The parabolic cylinder collector.

tage to purchasing a ready-made item. While the aperture area of our 32-ft reflector is about 804 ft², the area of the surface of the reflector is about 869 ft². If good quality plate glass mirror tiles could be purchased for about $3 per ft², the reflecting surface can be covered for about $2,600.

The possibility exists of obtaining surplus paraboloidal antennas of the proper size. For radar work, these antennas are usually covered with a metal mesh or grating for the reflecting surface. In the absence of a precise mold, the surface of the reflector could be tiled by hand with the 16,600 or so tiles, each of which could be adjusted optically for focus when bedded in the cement. This would add about 3000 lbs to the weight of the reflector if ¼-inch plate-glass mirrors were used. The increased wind resistance and the increased weight would probably still lie within the design limits of the reflector structure and the antenna mount. After all, these units are often designed to withstand 120-mph winds while covered with 1 inch of radial ice, which would generally choke the holes in the reflector surface. The overall weight of the antenna and mount thus treated will probably lie in the 5,000 to 6,000 lb range.

This solution is be far less expensive than trying to build the reflector and mount from scratch. The reflector would tend to look a bit like the ball hung in the Aragon Ballroom but it would be quite effective.

ALTERNATE FORMS

A somewhat less difficult reflector to build is the parabolic cylinder-type collector illustrated in Fig. 11-11. This type of reflector is parabolic in one direction and a straight line in the other. Because of this it focuses the rays of the sun into a line rather than a point. The boiler, therefore, can be constructed of tubing.

The parabolic cylinder is easier to build than a paraboloid of revolution for rather obvious reasons. It has the advantage that it needs aiming precisely in only one plane since a slight mis-aiming in the linear plane only tends to move the image endwise along the boiler.

If one pursues the same calculation performed for the parabolic cylinder of 32-ft width 24-ft height, one finds that the 50,000 BTU per ft^2 per hr rate would give a boiler tube only about 0.94 inch in diameter. The tolerances required for such a small target would be impossible at practical cost in a structure of this size. However, if the heat input rate were reduced to 2,500 BTU per ft^2 per hr, one would find that a boiler radius of 9.4 inches would result. The inset in the illustration depicts a boiler made of a series of standard tubes brazed together in a semicircle, enclosed in the glass tube radome. The reduction in heating rate would be accompanied by a similar reduction in flow rate through the boiler, if the same system pressure was to be maintained.

The low heating rate and the length of the boiler have some other problems in that a very substantial gradient in water temperature would exist if the flow were simply a straight line. The water at the exit end would be a great deal hotter than the water at the entering end. Some help could be obtained by making the water in the even numbered tubes enter at one end of the boiler and the water in the odd numbered tubes enter at the opposite end. The boiler would probably have to be segmented as well so that the water would enter and exit at a number of points along the structure.

A low transfer rate boiler has certain problems. In order to operate smoothly, a boiler should have the water enter and leave in water form. It should not flash off into steam inside the boiler tubes themselves. In a high transfer rate boiler, the high velocity of the water promotes turbulent flow through the tubes. This tends to scrub the tube walls and prevents the formation of steam films and bubbles. Since the specific heat of the steam is less than that of the water, these can give rise to local overheating of the tubes. The violent collapse of the bubbles can make the boiler noisy and can promote vibrations which fatigue the tubes. The use of the small tubes promotes turbulent flow which tends to minimize this action.

188

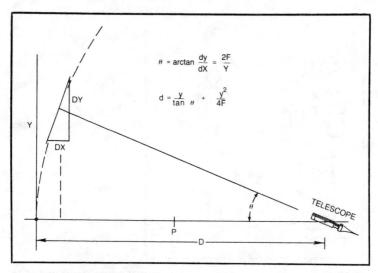

Fig. 11-12. Tile alignment.

Another advantage to the use of the small tubes is the fact that for a given pressure, the wall thickness of the tube may be decreased by about the ratio of the diameter. This means that there is less metal through which the heat must flow, and the unit therefore responds faster.

An advantage to the parabolic cylinder collector tiled with plate glass mirrors is the fact that there are far fewer tiles to be installed and adjusted since the tiles can be very long in the straight line plane. The arrangement shown in Fig. 11-12 is a technique for aligning the tiles. If the angle θ is calculated along with the distance, a surveyor's transit may be set up along the focal axis of the reflector. The tile is correctly aimed when the image in the mirror shows the transit precisely centered on its own crosshairs. This technique may also be used on paraboloids of revolution in which case the angle may have to be resolved into azimuth and elevation components for tiles not in the principal plains.

THE POWER TOWER

It was noted earlier that the cost of reflectors tend to rise faster than linearly with area. As a matter of fact, the cost seems to correlate with about a 1.3 power of area. This is due to the fact that as the reflector gets larger in area, it becomes higher and requires even more support and structure to hold its shape in wind and under the pull of gravity. Since the 2.24 kW plant is seen to be non-competitive with commercially available steam or oil-fired units on a

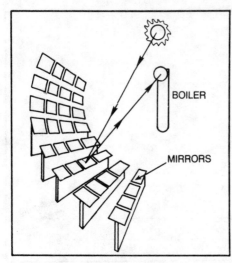

Fig. 11-13. The power tower.

cost per installed kV-A basis, a 1 megawatt commercial unit (446 times the power) would be nearly 2800 times as bad on a cost basis. It is obvious that some other technique must be found for solar energy collection if it is to succeed on a commercial scale.

One possibility has been suggested in the form of the *power tower*. Figure 11-13 shows this concept. A central tall tower which supports a solar boiler is mounted in the middle of a field with an array of small to medium sized mirrors which are individually mounted ringing the northern side of the tower or perhaps completely surrounding it in low latitudes. Each of these mirrors is individually gimbaled so that it can send its reflected rays to the boiler. In another refinement, the unit under construction and test by the Sandia Corporation has a small force transducer in the center of the back of each mirror. Pulling on this transducer deforms the square mirror, which is supported at the corners, into a paraboloid. A computer is required to continually calculate the aiming angles and the parabolic focal lengths required. These parameters are perpetually changing and must be continuously updated.

The power tower has the advantage that only the tower itself need be a high structure. Furthermore the mirrors, which tend to be about eight-feet square, are small repetitive structures which would lend themselves to mass-production economies. In initial tests with only a fraction of the mirrors in operation, a three-foot hole was burned in a few minutes through a one-inch thick sacrificial steel plate placed in the boiler location. Three major corporations are competitively testing boiler designs to be used with this collector. A nominal 1 MW rating has been specified for the unit.

Mounting and Tracking of Solar Collectors

12

From the previous discussion, it should be apparent that it will be necessary to cause the solar collector for an electric plant to track the sun in its motion across the sky. It is true that solar heating systems which do not track the sun are feasible. However, a solar steam plant requires a much greater concentration of heat than a solar heating system. For the latter, the only slightly concentrated solar flux is sufficient to warm the water. For the solar steam plant, though, the water must not only be warmed but boiled, which requires a concentration factor of at least 50 of the solar flux. The concentrator must be moved in order to have the concentrated flux impinge upon the boiler. The relatively high concentrations discussed in the previous chapter will generally require a tracking of the position of the sun to an accuracy on the order of a quarter of a degree.

The parallel between the solar collector and the optical or radio telescope was drawn earlier. A large number of different mountings have been developed for radio telescopes. The radio telescope is the more nearly applicable because of the size to which the instruments are built and because of the usually short focal length of the instruments. By comparison, the optical telescope has a very long focal length for a given size and this reflects itself in the mechanical design of the mounting.

For a great many years, astronomical instruments were constructed mainly with *equatorial* mountings. In the equatorial mounting, one axis of the instrument was aligned very precisely parallel to the axis of the earth. Rotation of the instrument about this axis was said to be in the *hour angle*. A second axis perpendicular to this was called the *declination* axis. Once a star was located in both declination and hour angle, the declination axis could be locked and the star tracked by motion of the instrument in hour angle only at the rate of 15° per sidereal hour. The equatorial mounted instrument has several distinct advantages:

- The mechanical drive system need be applied to only one axis of the mount.
- The position of the object could be read by use of setting circles applied to both axes with no mathematical correction. The declination axis reads declination directly and the hour angle axis reads hour angle.

The principal drawback to the use of the equatorial mounting is the fact that one of the shafts must be inclined with respect to the vertical for installations in all locations except for the north and south poles. Machinery as large as is required for a solar collector is much less expensive to build and install if the axes of the machine are either vertical or horizontal. For one thing, with an inclined axis machine, if there is any unbalance in any of the machinery it will tend to swing down to the lowest position. The drive for the machine will have to push the machine uphill until the equilibrium point is passed and will then have to hang on to prevent it from coasting downhill of its own accord.

By contrast, the *altazimuth* mount has one shaft vertical and the second shaft horizontal. The machinery is required to have a precise balance only about the *altitude* axis, and a small unbalance in the vertical *azimuth* axis will not encourage any self rotation effects.

Figure 12-1 shows a comparison of the two mounting types. It is usually desirable to have the two axes cross. In the case of the equatorial mounting, this means that the center of gravity of the collector and the declination drive does not pass through the center of the hour angle bearing. Thus, there is a considerable overturning moment on this bearing. By comparison, the altazimuth mount has the center of gravity of the rotated assembly directly on the axis of the azimuth bearing. At low latitudes, the fork of the equatorial mounting can be extended, and a second bearing placed upon the top of a pier at the northern end of the hour angle axis. However, at mid latitudes, the altazimuth mount has a distinct structural advantage.

The principal drawbacks to the use of the altazimuth mount was that all of the angles had to be calculated from the positions of the azimuth and elevation circles, the drive motion had to be applied simultaneously to both axes, and the drive was non-uniform with

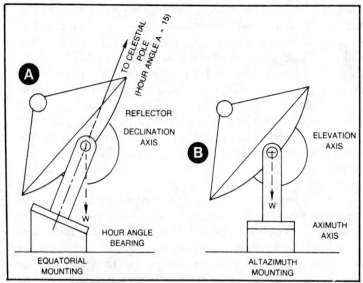

Fig. 12-1. A comparison of the two types of mountings for solar collectors, the equatorial at A and the altazimuth at B.

time. Since the development of the minicomputer, many of these objections have disappeared, because the minicomputer can be programmed to give a running translation of the declination and hour angle commands into azimuth and elevation angle commands with any required degree of precision. Accordingly, the use of the altazimuth type mount has become increasingly popular for astronomical instruments. The savings in the mechanical construction more than offset the costs of computerizing the drive.

Probably the cheapest way to obtain a suitable altazimuth mount for a solar collector is through the surplus market. Most tracking-type radars employed a mount of this type and are directly adaptable to this application. Another possibility lies in the use of a five-inch naval gun mount. In the early days of radio astronomy during the 1950's, the five-inch gun mount was the most popular choice for a radio telescope mounting. A number of these gun mounts are probably still available from scrapped WW II destroyers and cruisers. The mount is extremely strong and stable and quite capable of supporting reflectors up to 50 feet in diameter in wind and ice.

A HOMEBREW MOUNT

If one of the sophisticated antenna or gun mounts is not available through the surplus route, a suitable mount can be constructed of wood and available hardware along the basic pattern shown in Fig. 12-2. In this case we see an octagonal frame with a series of four trailer wheels mounted in a circle. A surplus truck rear axle half is used as a center pivot, but all of the weight is carried by the trailer wheels. The wheels simply roll in a circle upon a cast concrete pad, and the truck axle is imbedded in the center of the pad. The main frames would probably look neater if they were mortised to lie in the same plane. However, this tends to reduce the strength by about one-half and the stiffness by a factor of four. Therefore, it is probably better to use them uncut. The design carries the weight around the periphery, and all of the overturning moments are carried directly to the trailer wheels.

It would probably be good construction practice to first cast the circular pad for the wheels to run on, leaving an open excavation for the truck axle. The truck axle could then be attached to the finished turntable leaving plenty of shim room to make up for variation in the tire pressure. Once the table was centered on the track, the axle could be cemented in place, thereby assuring that there was no

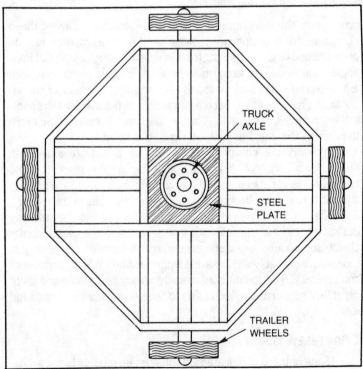

Fig. 12-2. A simple homebrew azimuth turntable.

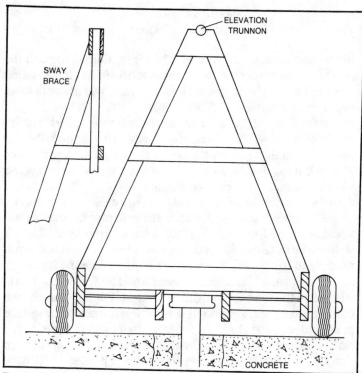

Fig. 12-3. The elevation axis.

interference with the truck axle. After the cement hardens, the axle can be loosened from the steel plate and the shims removed so that there is about an inch of vertical "float" and the axle carries no vertical load. If this is not done, the turntable will be almost certain to bind in some positions and for some conditions of tire inflation. If the steel plate is attached from the top, the servicing of the axle will be facilitated, and it will be possible to jack up one wheel for removal in case of a flat tire.

The view of Fig. 12-3 indicates the fork and trunnion construction. This unit relies upon an "A" frame type of construction to carry the weight and overturning moment down to the trailer wheels rather directly. The sway braces are a very necessary part of the construction since the unit will want to be turned toward the sun and the wind may approach from any angle.

Wind force upon this structure can be very considerable, as noted in the section on windmills. If we consider the device as having the same resistance as a flat plate, the wind will deliver a pressure of:

$$P = 0.0042\ Va^2\ lb/ft^2$$
$$V_a = \text{wind velocity in mph}$$

In a 90 mph wind, this force is 20.1 lbs per ft². Considering only the area of a 32-foot collector, this amounts to 16,165 lbs. If we assume a height for the elevation trunnion of 20 feet, this amounts to an overturning moment of 323,300 ft-lbs.

If the radius of the turntable out to the tires is 12 feet, and the weight of the turntable and collector is assumed to be 8000 lbs, the deadweight at the center of the table will supply a righting moment of 96,000 ft-lbs. If a ballast of 4800 lbs were applied at each of the wheels, the unit would be in equilibrium. However, the ballast and deadweight would bring the wheel loads up to 6800 lb per wheel.

A far more practical approach would be to consider that the unit would be stowed with the reflector pointed vertically at all winds above some more modest level. If we were to design for operation in winds up to 40 mph, the maximum pressure would be 4.7 lbs per ft², and the maximum overturning moment would be 3780 ft-lbs. This is well within the deadweight righting moment. For higher winds the collector would be stowed pointing vertically to minimize the surface presented to the wind and the reflector itself could be guyed to the earth using pre-built anchors installed at erection time. In northern climates, the reflector may have to be stowed pointed horizontally to minimize snow loads.

Note that the sketches in the illustrations have left out a number of items, such as gussets and bearings to captivate the elevation trunnion, for the sake of clarity. These items *cannot* be left out in the finished unit. The dimensions of the turntable have also been left out since these will be a function of the collector size. A device of this sort will generally be quite stable if the radius to the tires is about 3/4 of the radius of the reflector.

RANGE OF ANGLES

The construction of the antenna mount is obviously influenced by the range of angles through which the reflector must be swung. Figure 12-4 shows the situation for an operation in the northern hemisphere. The sun is always highest at local noon (standard time) and the elevation angle above the local horizon is $90° - L + D$, with L being the north latitude of the site and D being the declination of the sun at the time. Actually, for reasons of stowage noted earlier, it is usually advisable to be able to point the collector straight up. The limitation is therefore placed by the stowage condition rather than the range of sun angles.

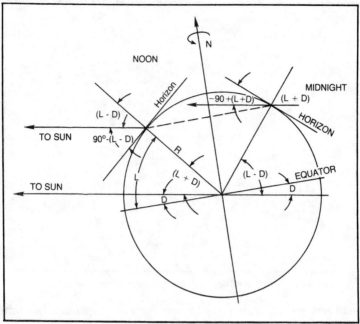

Fig. 12-4. Sun elevation angles.

The range of azimuth angles is also limited. The azimuth of the sun is given by the formula:

$$\cos Z = \frac{\sin D - \sin L \quad \sin h}{\cos L \quad \cos h}$$

where h is the true height of the center of the sun at the moment.

As a practical matter, it is not possible to collect any significant amount of solar energy with the altitude of the sun much less than 5° above the horizon. This offers some latitude in the construction of the steam collecting system: Full rotary joints are not needed because it is not really necessary to rotate the collector through a full revolution. The curve of Fig. 12-5 furnishes an approximate table of sun declination versus day of the year.

TRACKING THE SUN

Two fundamental approaches may be taken to track the sun. The position of the sun is well enough known so that the collector can be precisely calibrated and the reflector driven to track the sun with no sensing. However, this implies a knowledge of the direction in which the collector is pointing to a very high level of accuracy. The pointing angle is affected by flexure of the mount and a variety of other errors. In order for the preset tracking arrangement to be

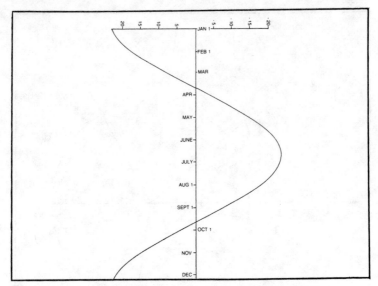

Fig. 12-5. Approximate declination of the sun as a function of date.

sufficiently accurate, it would be necessary to compensate for each of these.

A far more practical mechanism for tracking the sun is to employ active sensing. At any time that the sun is visible enough to provide any significant amount of solar energy, it is a relatively simple matter to sense the pointing angle of the collector and drive the unit to an optimum pointing angle.

Figure 12-6 illustrates one such technique. The shadow of the boiler is allowed to fall on a pair of solar cells so that when the collector is properly pointed, each cell is about half covered. The output of the two cells is then compared and used to derive a signal to drive the collector in one or the other direction to restore the signal balance for the two cells. By the use of a pair of suitable baffles or fences, the servo can be made to function even when the sun is so far off aim that the shadow of the feed misses both cells. The design of a servo of this type is somewhat beyond the scope of this text. The reader is directed to TAB book No. 929, *Solid-State Motor Controls,* by the author.

It should be noted that the position of the shadow of the boiler upon the reflector is a very sensitive measure of tracking. Therefore, the system can be made to track with very suitable accuracy. Because the servo is actually sensing the final parameter to be optimized, the effects of errors and flexure in the mount and the collector itself are essentially eliminated.

SYSTEM STABILITY

The simplified servo drive of Fig. 12-6 is, of course, far too simple to function effectively in a continuously operating solar tracking system. In the first place, the system must have some significant amount of deadband in order to prevent the system from continuously hunting or jittering. This means that there must be some small angle where the servo is completely shut off and does not attempt to drive the reflector in either direction.

As has been noted earlier, the reflector and altazimuth mount assembly will be a rather large and heavy structure. The motor and drivetrain used to drive the assembly to track the sun will generally have a very large gear reduction. For example, suppose that the wheels have a *rolling radius* of one foot (rolling radius is the effective radius which describes the forward motion of the wheel under rated load and at rated tire pressure, and it may be found by measuring the forward progress of the wheel in one complete revolution and dividing by 2π). If we assume that the radius from the azimuth axis out to the center of the wheels is 12 feet, then the circumference of the azimuth circle is $2\pi \times 12$ feet or 75.4 feet. With a rolling radius of one foot, the wheels will advance 6.28 feet per revolution or will require 12 revolutions to complete 1 circuit of the azimuth circle. Stated differently, the azimuth angle of the reflector will change by 30° with each revolution of the wheels. At noon in mid-latitudes, the required azimuth rate for the reflector is about 15° per hour. Therefore, the wheels will be turning at the rather slow rate of a half revolution per hour.

Let us suppose that we wish to drive the reflector in azimuth with an ordinary induction motor which turns at 1750 rpm. We would probably like to have a capacitor-run type motor because of the ease with which reverse operation can be obtained. If we chose to have the motor capable of slewing the collector assembly at a rate of 45

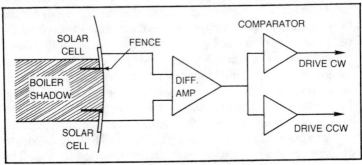

Fig. 12-6. A simplified servo drive system for solar tracking.

degrees per hour, our average motor speed would be about 583 rpm, and we would require a gear reduction ratio of about 583 × 60 min per hr × 2 hr per rev = 70,000:1 between the motor shaft and the wheels.

There are several facets to the use of such a large gear reduction. First and foremost is the fact that it will not be necessary to use a very large motor to drive the collector because the huge gear reduction will multiply the torque of even a small motor up to levels adequate to move the reflector, provided that it will roll freely.

A second and less desirable facet of the very large reduction comes from the multiplication of any play or looseness in the gear train. If we presume that a looseness or "slop" in the gear reduction is to be held down to a total swing in aiming of the reflector of a quarter of a degree, this would be multiplied by 12 to obtain the looseness in a given wheel and again by 70,000 to obtain the looseness measured at the motor end of the train. This amounts to 210,000 degrees of rotation of the motor or 583 revolutions. If the motor were driving the unit at standard rate in one direction, and the unit stopped and the motor then reversed to cause the unit to back up, the motor would run for *20 seconds* in the wrong direction before it had taken the slop out of gear train and the collector began to move in the opposite direction!

It is possible to obtain a 70,000:1 reduction using a small 100:1 worm gear speed reducer driving a second unit with 100 times the torque rating and a second 100:1 reduction and perhaps obtain the final 7:1 reduction using a timing belt. But this arrangement would not prove too satisfactory in practice for a solar tracker since worm drive reduction units tend to have a very large lost motion or play or slop. A far superior arrangement would employ one of the harmonic drive units which can be obtained commercially with reductions in excess of 1000:1. These units tend to have inherently less lost motion than a worm drive. An equivalent advantage exists for the use of a timing gear drive compared to a chain at the wheel end of the reduction. The timing belt arrangement can be made with relaxed tolerances and a takeup pulley employed to take out the majority of the lost motion. For truly satisfactory solar tracking it is probably necessary that the drive reduction gear train be tight enough so that the total lost-motion at the motor end represent no more than about five seconds running time at the motor end at operating speed.

In a practical system of this sort, it is not usually a good idea to attempt to achieve proportional control over the motor. The simplest and most effective servos for a drive of this sort usually turn the motor completely on or completely off. The motion of the reflector will usually represent a series of very small jogs. If the operation of the system is otherwise smooth, the motor will cycle on and off without the motor ever coming to a complete stop.

PLAY AND SWAY

While it is necessary to have the drive capable of slewing at a rate some multiple of the maximum tracking rate, it is not advisable to have the overall system too "zippy." In the normal operation of a tracker such as the solar tracker, the target will frequently be lost behind a cloud or otherwise obscured. During this period tracking errors will have the opportunity to build up. When the sun reappears, it is necessary that the collector be able to go significantly faster than the standard tracking rate to give it a chance to catch up. This catch-up rate must be so much faster than the normal tracking rate that the image of the sun or a fraction of the image will not remain focused on the non-cooled portions of the feed structure. At the solar concentrations required for a good solar boiler design, the heat of the image of the sun focused on the boiler support struts could be adequate to cause severe overheating, distortion, and in the extreme case, melting of the structure. The boiler is equipped to take the heat away whereas the support structure is not.

On the other hand, a system capable of providing accelerations which are too rapid is not desirable since it can cause severe mechanical stress in the collector and support. The boiler, reflector (or refractor) frame, and the elevation mount will all have a certain amount of elasticity. This can give rise to system instabilities in several ways.

For example, consider the arrangement shown in Fig. 12-6. Suppose that the unit lost the sun behind a cloud for a period of several minutes and that thereafter, the sun emerged and the top cell were completely exposed and the lower cell completely shadowed. The servo would immediately begin driving the boiler upward to equalize the difference, probably at top speed. When the shadow was again equalized, the motor power would be removed and the motion would tend to slow down. However, the system has some substantial amount of inertia, and there must always be a certain amount of flexibility in the azimuth suspension, the support structure, the collector itself and the feed structure or boiler support. Even if the motor were to stop instantaneously, the flexibility in these components would permit the entire assembly to swing somewhat past the mark.

For example, let us consider a unit with a 10-foot focal length, where the sensing solar cells are 2 inches in diameter. If the total overswing at the feed were as much as two inches (0.95°), the upper cell would be completely shadowed (briefly), and the drive motor would be turned on in the reverse direction. At just about the same time that the collector would be trying to swing back toward the neutral position, the drive motor would have taken the slack out of the system and would be driving the collector full speed in the downward direction. This drive could cause the feed to whip to the low position again where the motor would be excited to drive the entire system upward and the system could go into a series of oscillations which might easily build up to the point where some portion of the structure would mechanically fail.

The calculation of the actual spring constant and moment of inertia of a structure as complex and large as a working solar collector is an enormously complex problem which would tax the resources of a fair sized mechanical engineering group and is probably completely out of the reach of most do-it-yourselfers. For this reason, an a priori prediction of the criteria for stability of a given system is probably not feasible. However, this does not mean that a homebrew system cannot be made to work. It simply means that it will probably be necessary to experimentally determine some of the parameters of the system empirically, after the construction of the collector. Certain rules-of-thumb can also be applied to determine whether the system will be stable.

As an example, let us turn back to our ten-foot focal length system. If we presume that the system pointing accuracy must be maintained within 0.25° and that the effective center of swing of the system was at the vertex of the reflector (or at the surface of the boiler for a refracting system) we may perform a few simple calculations to determine the required stiffness of the system. At a radius of 10 feet, an angle of 0.25° describes an arc of 0.52 inches. If we arrange the deadband for our comparator such that the motor cuts off while there is still a 0.52-inch gap between the center of the feed shadow and the axis of the system, we will have arranged the system to have the maximum amount of deadband gap consistent with the tracking accuracy required. The system will shut down and coast as soon as possible (for a simple system).

There is a rule-of-thumb which can be applied to simple on-off or "bang-bang" servo systems which says that the system will generally be stable if it coasts to a velocity less than half of the initial velocity when it hits the far side of the deadband. If we assume that the swaying motion of the entire collector system is sinusoidal with time, then this can be shown to correspond to an overswing of no more than 15 percent of the deadband offset. Stated differently, if the motor shuts off when the shadow reaches the point 0.52 inches *before* the axis, and if the overswing does not permit the shadow to travel more than 0.52 × 1.15 inches *beyond* the axis, the system will probably be stable.

It should be noted that this rule applies only if the motor responds instantaneously. Since we have seen that this is apt not to be the case because of lost motion in the drive, the actual tolerance must be somewhat smaller.

One of the assumptions in the previous discussion should be considered further, namely the center of radius of the overshoot swing. Figure 12-7 contains the assumption that the entire collector mount and assembly are perfectly rigid and that all of the overswing is due to the flexing of the tires in the azimuth mount. Here it may be seen that the effective center of the overshoot is very near the center of the azimuth mount. The overshoot positions are shown dotted. In actual practice, there will be a number of sources of flexure in the system and the actual travel of the collector may not be in an arc of a circle. When conducting overshoot tests, the actual travel of the image is the most significant item and should be used as the criterion for the test.

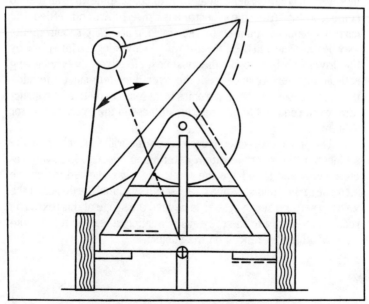

Fig. 12-7. The effective center of overswing. The assumption is made that the entire collector system is perfectly rigid and all overswing comes from the flexure of the tires in the azimuth mount.

203

Obviously, the distance that the overshoot tends to cover is a function of the speed at which the reflector was slewing. The slower the reflector was travelling when it entered the deadband, the less likely it is to coast too far. The restriction on the maximum slewing speed is therefore an effective way to contribute to the stability of the overall system. If, in the course of initial tests of the system, the overswing is found to be excessive, the system must either be stiffened or the slew rate reduced.

It should be noted at this point that inertia effects are not the only ones which can cause the system to overswing. The wind can also contribute to this problem. In initial tests, if the wind-induced sway is found to exceed more than about half of the pointing tolerance of 0.25°, then the system will either have to be stiffened or the operation of the system on windy days restricted.

VARIABLE DEADBAND SYSTEMS

One obvious way to reduce the tendency of the system to oscillate or hunt is to make the deadband a function of the speed of the system. If the actual attained speed of the system is measured and used to adjust the deadband, the restriction on the slewing speed can be raised somewhat. The unit will then shut down the motor somewhat earlier and allow it to coast somewhat further to come to a stop. After the collector has slowed down and settled, the narrow deadband required by the tracking accuracy requirements may be restored and the collector will be slowly jogged into place by the drive. This is probably the most cost effective way of achieving satisfactory performance from the overall system since the additional complexity in the electronics is far less costly than the structural work required to provide the stiffness in the overall collector system.

The most cost effective compromise will probably find the stiffness of the overall system determined by the wind sway requirements and then have the remaining sway allowed for by the increase in the sophistication of the servo system. The design of the velocity sensing servo system is beyond the scope of this text. The reader is referred to the previously mentioned text on motor controls or other references on the subject.

13

Solar Collector
Construction

At the time of this writing there are a few firms attempting to get into the business of constructing solar collectors suitable for steam production. However, standard units with published prices are not yet in production in the 20- to 35-foot diameters required for production of the amounts of electricity required to fully satisfy the needs of the typical American home. It is possible to obtain quotations for units, but these generally are furnished only when a full set of specifications is submitted and a genuine interest in buying is evidenced.

If we consider satellite terminals of the same collector size equipped with mounting and drive, we tend to find that a 30-foot diameter reflector will run in the price range of $50,000 to $150,000. It should be noted that such reflectors are not directly suitable for solar collection since they generally do not have an optically mirrored surface. In general the tolerances on the surface contour are in the same region as required by the faceted reflector described in Chapter 11. The rework of a satellite terminal or radar antenna would be cost-effective only if such units were obtainable as surplus at distress prices. For these reasons, it seems likely that, at least for some time to come, it will be necessary for the home electrifier to build his own solar collector.

Compared to a high-speed windmill of large size or a significant water impoundment, a solar collector presents few dangers to persons not in the immediate proximity of the unit. The dangers of boiler explosion have been noted earlier and the dangers of steam leakage will be discussed subsequently. However, for the collector itself, the principal danger is that the unit might blow down or collapse under the weight of a snow load. Neither of these dangers extends very far from the immediate vicinity of the collector. For this reason, the design and fabrication of a solar collector is probably not outside of the capability of the amateur designer and builder provided he lives in an area where failure of the unit will yield damage confined to his own property.

In many respects, the construction of a solar collector has requirements similar to the construction of a small aircraft. The collector must be relatively precise and regular and exceedingly stiff for a structure of the size and weight. It is furthermore usually desirable to make the structure as light as possible. The collector structure must be supported by the elevation gimbals and be moved by the elevation drive. The elevation axis fork is then placed upon the azimuth motion assembly which must rigidly support and propel both units. An extra ounce in the collector proper will add a half pound to the weight of the fork which will add three pounds to the azimuth drive! Therefore, it is not unusual to find that the construction of large collectors will follow aircraft techniques.

REFLECTOR CONSTRUCTION

The typical large reflector is generally constructed with a back frame to give the required rigidity and a precision front surface which is determined by a mold of some sort. The backframe is often constructed from aluminum or steel tubing in an elaborate trusswork to provide the features of rigidity combined with lightness. The basic attempt is to divide the structure into a series of cells fabricated from triangles.

Figure 13-1 shows the view of a reflector of this construction from along the focal axis. It may be seen that this reflector has been divided into a central hexagonal core and a series of six panels. For economy of manufacture, the panels are identical. The radius of the central core is 40 percent of the radius of the reflector itself. As the ratio of focal length to diameter grows larger than the value of 0.3 shown in the example, it is usually wiser to reduce the ratio of central core to overall diameter.

Figure 13-2 shows the structure of the backframe in diagramatic form. This view is shown along a plane going through the vertex and one of the panel dividing points. Points A and B represent the point of attachment of the panel to the central core. If the panel is loaded by a wind blowing in the direction from the focus to the vertex, it may be seen that stringer LKHEB is in compression and

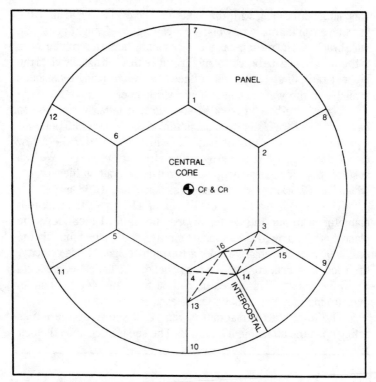

Fig. 13-1. Subdivision of the reflector along the focal axis.

stringer LJFDA is in tension. Since the stringers are relatively slender with respect to their length, it is necessary that something be done to keep the stringer that is in compression from buckling. The spacers DE, FH, and JK serve to hold the stringers apart and the diagonals AE, EF, and FK serve to turn the trapezoidal areas made up of the stringers and the spacers rigid. Since a wind-loading from the rear of the reflector is equally likely, it is necessary that both stringers be capable of operating in compression.

The advantage of the panel type construction is the fact that the mold may be made much smaller than would be required if the reflector were made in one piece, and much less indoor assembly space is required. In addition to this, there is the advantage that two men can handle the sections, whereas the finished reflector will certainly require the services of a crane, and a rather large one at that.

On a structure such as this, the builder has a variety of options for the truss construction. The trusses could be made of thin-wall steel tubing with the joints welded or brazed or somewhat thicker

207

wall aluminum tubing with the joints heliarc or TIG welded or the structure can be made of wood with glue, gussets, light nails, and wire bracing. The most popular choice among home aircraft builders is the wood and wire braced structure since this will usually yield the highest ratio of strength and stiffness to weight using techniques available in the well-equipped home workshop.

For a wire-braced wooden structure, it is fairly common to have a ratio of truss depth to diameter on the order of 25 percent. Operating on these proportions, it is instructive to make a few calculations regarding the size of the stringers required. We will assume that the construction material is aircraft grade spruce, which has a working stress in compression of about 768 lbs per in^2.

The panel has a total frontal area of 116.3 feet for a 32-foot diameter dish, and the center of pressure is 5.51 feet above the plane described by AB. Assuming that the wind "fastest mile" in the area is 90 mph, we shall consider a 100 mph design. At this velocity, the wind pressure, considering the panel to be a flat plate, amounts to 42 lbs per ft^2. This gives us a total force of 2442 lbs and an overturning moment of 3364 lbs at each rib.

The force vector diagram of Fig. 13-3 shows us the forces acting upon the attachment bolt at B. The shear force upon the bolt

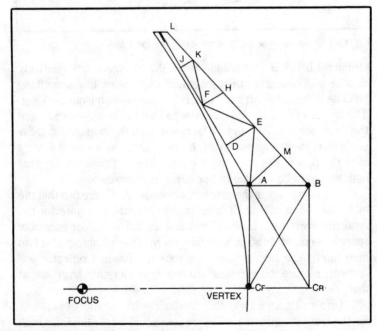

Fig. 13-2. Back truss construction.

208

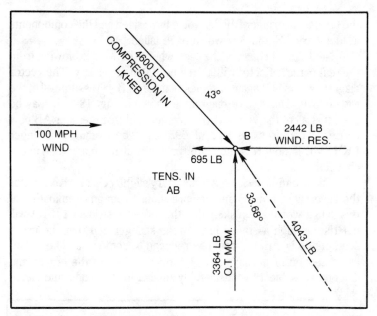

Fig. 13-3. Forces acting upon clevis B.

is 4600 lbs, which is equal to the compressive force in stringer LKHEMB. For this compressive force at 768 lbs per in^2, we would require about 6 in^2 for the stringer, which would probably be in the form of a 2×3 (finished) member. If we take the working shear stress of an automotive quality bolt to be 10,000 PSI we find that the bolt would require a cross section of about 0.52 inches. We would probably use a 9/16-inch bolt (this assumes a double-shear clevis arrangement).

You will note that in this analysis, it was assumed that the entire resistance to the wind force which would act to slide the panel straight back is furnished by the downwind clevis. Furthermore, this implies a moderate amount of tension in the member AB. If the actual tolerances in the fit of the bolts and the clevises had been such that the upwind clevis could share some of the wind resistance load, the tension would have risen in member AB. However, the compressive stress in the stringer LKHEMB would have remained constant.

Going now to clevis A, we observe in the force vector diagram of Fig. 13-4 that both the total shear foreces on the clevis bolt and the tensile and compressive stresses in the members intersecting at A are somewhat smaller than those at B. If A had shared somewhat more in the assumption of the sliding component of wind resistance, the tension in stringer LJFDA would have risen and the components in members AE and AM would have fallen.

Next, let us consider the case where the wind is blowing from the rear of the reflector with the same 100 mph velocity. The vector diagram at clevis B again assumes that clevis B will assume all of the sliding load. This vector diagram is shown in Fig. 13-5. It may be seen that all of the vectors are reversed and that member AB now carries a compressive load of 695 lbs. The loading in stringer LKHEMB is now reversed into tension, but the magnitude is unchanged.

The situation at clevis A is also very slightly changed except for the reversal of tension and compression. From an examination of this data it would seem likely that the forces in stringer LJFDA will not rise as high as those in the rear stringer and that the cross section of this stringer could be reduced to some value like 4.7 in^2 cross section or perhaps 2×2.5 inches. Each of the braces and diagonals is subject to considerably smaller loading and could there-

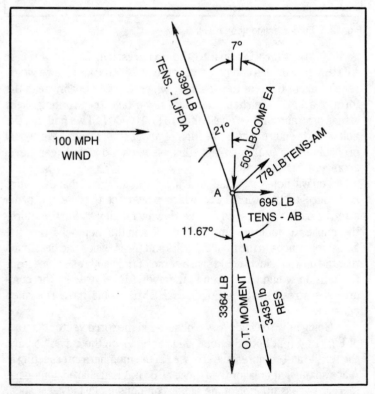

Fig. 13-4. Forces acting upon clevis A.

210

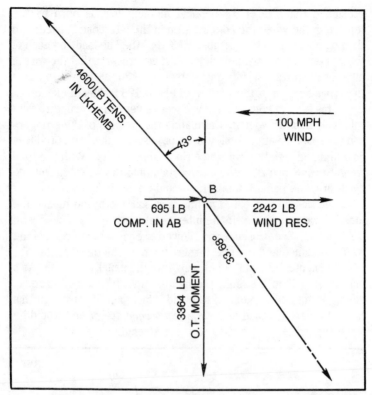

Fig. 13-5. Forces acting upon clevis B with the wind from the back of the panel.

fore also be reduced. Also the bolt at clevis A could be smaller than that at clevis B. However, the weight savings here would be negligible.

It is fairly obvious that the rib structure would be fairly stiff and rigid. Yet we have postulated a structure in which the entire loading from the panel is carried by four clevises at the bottom edges of the panel. The actual load is distributed across the panel and there would be a tendency for the panel to sag in the center if the panel were not infinitely stiff, in the direction of the circumferential members. The panel would also be subject to racking stresses in certain positions. Referring back to Fig. 13-1, we see that panel 3, 4, 10, 9 has been shown with an intercostal rib which is between the mounting points at vertices 3 and 4 of the central core. The criss-cross dotted lines represent wire braces added to stiffen the panel and to prevent racking.

Figure 13-7 shows the effect of the wire bracing to hold up the intercostal between points 13, 14 and 15 at section EF of the rib. It is presumed that the entire bending moment load is born by the tension in the wire, the compression in the intercostal spacer and the compression in the member E13, 14, 15. The loads on each of these members is relatively modest. If we presume that the wire is music wire with a working tensile strength of 30.000 lbs per in^2, then the cross section of the wire would be 0.128 inches in diameter.

This type of construction will yield a very rigid structure with relatively minimal weight. Calculating the weights of the stringers and the spacers with due allowance for gussets, glue, etc. produces an estimated weight for a single rib of about 60 lbs. With the total weight of the two ribs, the intercostals, and the wire bracing and hardware, the panel framework should be about 250 lbs.

The design presented in this discussion is by no means the most efficient possible. If the intercostal were also attached with clevises to the central core and allowed to bear some portion of the wind resistance and the overturning moment, all three of the ribs could be lightened and the net weight of the panel would fall. As a matter of fact, in the limit, the lightest structure would have, in effect, an infinite number of ribs and attachment clevises. In this case, the design would become a *monocoque design* and would be about the lightest achievable with the strength.

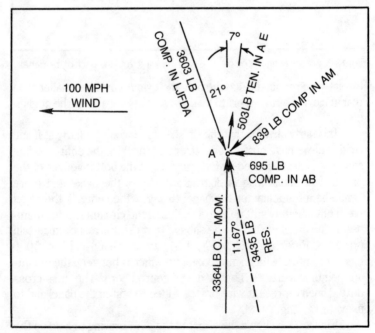

Fig. 13-6. Forces acting upon clevis A with the wind from the back of the panel.

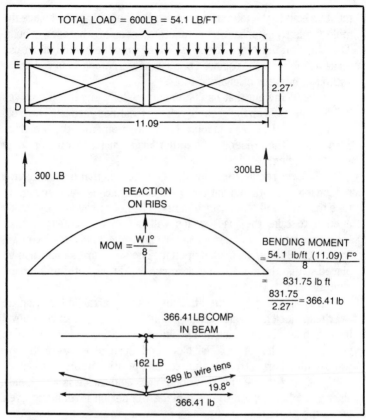

Fig. 13-7. Wire bracing of the intercostal.

In actual engineering practise, skin/foam/skin or honeycomb structures do come very close to the theoretical limit predictable for a true monocoque structure. For molded fiberglass structures with a honeycomb or foam filler, the strength to weight ratios are very high. Unfortunately, these structures are not very easy for the home builder to produce, and they generally require a substantial amount of tooling and no little amount of fabrication experience. It is expected that if commercial demand were to arise for solar collectors these techniques would be very commonly employed in their construction.

FABRICATING THE MOLD

For a shaped body such as a parabolic reflector and particularly for a paraboloid of revolution, it is nearly imperative that a mold be available for assembly and surface molding and so on. The mold

213

must be built to the tolerances required of the final product since the mold tends to establish the actual tolerances of whatever is built. The question therefore arises regarding how the home builder can make a mold with the required precision for a complex curved shape without a number of gigantic tools.

One technique that has been successfully employed involves building the mold on the ground with the focal axis vertical. The mold must generally be built indoors for protection from the weather. This implies the presence of a rather large barn or garage or some similar building.

The first requirement for this structure is that it must have a reasonably smooth and flat concrete floor which is thick enough to take the bolts and other attachments. The floor must also be strong enough to take the loads which will be placed upon it without sagging, settling, or cracking. Before the project is finished you will have 20 to 30,000 lbs loaded upon it. The structure should also be placed so that the mold floor will remain dry during both summer and winter.

It is also handy if the structure can be arranged so that an overhead block and tackle arrangement can be rigged since you will have to be carefully lifting and positioning articles which will weigh upward of 500 lbs. A triple fall block and tackle using 3/4-inch rope will permit a fairly strong man to lift and position items up to about 1000 lbs singlehandedly. A second block and tackle is also handy since a double pick on the workpiece will permit level lifting even when the exact center of gravity is not known.

THE CENTER PIVOT

The first step in the construction is to bore a hole through the floor and place the center pivot. This center pivot is a pipe upon which we shall swing the cutting apparatus that will shape the mold. The center pivot should be sturdy enough to bear a considerable load. The pivot could be made from ordinary plumbing pipe with the upper section turned in a lathe to assure roundness and smoothness or it could be thick-wall steel tubing stock purchased from a local steel supplier. Our mold will eventually be a keyhole shaped structure with the round part slightly larger than the central core and the slot portion slightly larger than the panel. The pivot should be located so that you can walk freely around all sides of the mold.

Figure 13-8 shows the general layout of the structure which shall be eventually constructed. In the layout of the workspace, it should be remembered that it will be necessary to machine the core area on both sides of center. Therefore, the building must have a

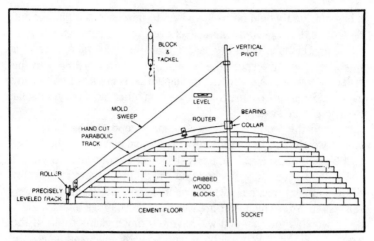

Fig. 13-8. The mold.

floor area large enough to permit the mold sweep to be walked in a full circle about the pivot.

In the first step, the socket for the pivot should be accurately located and cemented in with the pivot as precisely vertical as possible. The pivot itself should have no more than 0.015 inches of bow along the length, and the pivot should be vertical to within 0.015 inches. In order to assure these tolerances, it may be necessary to guy the pivot at the top with a series of cables so that the requisite vertical position can be adjusted and maintained. In this matter, the use of four guys rather than three will make for much easier leveling. Two people on opposite sides of the mold can slack off and take up on the turnbuckles to level the plane of the guys. The two people then walk 90° around the mold and level the guys in the other plane. A little interaction will exist between the sets, and only a few repeats to "touch-up" will suffice to make the pivot precisely plumb.

The next step in the construction entails the accumulation of a substantial amount of good, dry hardwood. Nearly any wood could be used, but maple is probably the best choice in terms of hardness and stability. It is most important that this wood be aged and dry and that a single type of wood—preferably from the same tree—be used in the construction. The wood can be in random lengths and does not need to be furniture grade, but segments with knots and twisted grain are to be avoided. In rural areas, it is not too difficult to find a farm sawmill where the local farmers will cut up trees for sale as construction grade lumber. The wood should have air dried for at least a year and a half with the last six months spent in an environment and humidity similar to the environment within your shelter. If

215

it has not, you would be well advised to purchase the lumber and allow it to dry, stacked with spacers between layers, within your building. This is intended to assure that the mold will not warp or split after you have spent the time and effort to machine it to the reflector contour. A six-month drying period is not a bad idea in any event and you can use the time to work on other facets of your solar electric power system.

Once the wood is duly seasoned, a good plan is to rent or otherwise obtain a large thickness planer and plane all of the blocks to the same thickness. The exact thickness is not too important. If you study your supply of wood you will find that there is probably some optimum thickness which will maximize the yield. Thicknesses less than 1.75 inches or greater than 3 inches are best avoided. If the wood is too thin you will have too much gluing to do. If the wood is too thick, you will have trouble with voids and gaps.

In the course of the thickness planing, the convex side of a bow should be removed first to obtain maximum straightness. The wood should be planed on all four sides and the ends trimmed off square with a radial saw.

The next step in the procedure is to lay down a sheet of 6-mil polyethylene on the cement to seal the moisture. This step is important and should *not* be neglected. The wooden billets are next laid upon the floor and the construction of the mold blank begun. Alternate layers of billets are laid cross-grain to one another after painting the area with a good grade of casein glue. It is a good idea to test the casein glue with your water. With some types of softened water, certain casein glues will not form a brown, sticky syrup but rather a gritty, malted-milk color mix with very little holding power. Before starting the layup it is advisable to test glue a few samples to make sure that your glue will work with your water. If problems are experienced, you can easily purchase distilled water for glue mixing.

The mold is simply laid up like an igloo with the layers and adjacent billets glued together. At the end of any day's work, a sheet of polyethylene can be thrown over the pile, and the pile can be weighted down with 50- or 100-lb sandbags. It will probably be a good idea to have a dozen or a dozen and a half sandbags around for this purpose. The best type of sandbags are the military cloth type rather than the commercial paper bags in which sand is usually sold. The bags should not be tightly filled so that they are stiff. Instead, they should be limp so that they will conform to the shape of the pile.

Eventually your igloo will begin to take the shape of the mold. Precision is not particularly important at this point. You should only strive to get the shape and size of the pile such that you can machine

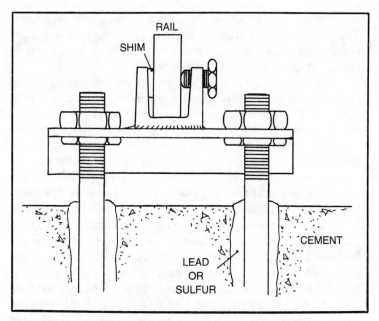

Fig. 13-9. The precision track mount.

the mold out of it. A little care in laying up the pile initially can save some machining time later. Once the pile is complete, it is advisable to give the entire thing a month or so to cure before starting any machining on it.

Once the pile is in place (and definitely not before) you can begin laying in the level track. This track is most easily constructed by using precision ground steel bar stock about 2 × 3/8 inches. This stock can be purchased from a steel supply house in lengths of three feet. The circular track will be approximated by a polygon with sides equal to the bar lengths. Locations for the track mounts can be laid out by a swing wire from the pivot post.

Figure 13-9 shows a fairly simple track clamp arrangement. A short length of channel is brazed or welded to the top of a length of angle iron. The latter piece is drilled to accept two bolts sunk into the floor and held in place with a casting of sulfur or lead poured into the holes drilled in the concrete. These will hold about equally well. The sulfur is cheaper and requires less heat to melt, but it is also more obnoxious and the fumes should not be inhaled. In either case, do not pour the molten material into the holes if there is any significant amount of moisture present; it will explode and send the molten material back out into your face. Always wear a plastic face shield while pouring.

The rail is firmly clamped between the shim and the setscrews. The shim is required in order that the rail will clamp in an upright position since the inside of the channel will always have a certain amount of draft or tilt to the inside wall.

The height of the unit can be precisely set by loosening the top nuts and running the bottom nuts up or down as required in order to level the rail. A good builder's or surveyor's level can be used to set the track sections level to within about 0.015 inches. After this process, a precision millwright's level can be used to fine level the track to about .002 inches. A bump at the joint between the rails can be avoided by having the ends of the rails overlap slightly in the polygon. The end of the mold sweep will be guided by this track. Therefore, it is important that the track be as level as possible.

The mold sweep is shown without much detail to simplify the picture. This has to be a sturdy frame equivalent in rigidity and lightness to the ribs but larger since it must reach all the way to the center of the reflector. A trussed construction similar to the ribs is probably a good choice. This unit can be layed out on the open floor area adjacent to the mold and assembled using the techniques to be described for the ribs. The hand-cut parabolic track can be a sheet of 12 or 16 gauge aluminum. This track can be installed before cutting to the ribs.

The parabolic curve is laid out on the aluminum sheet by carefully plotting out points about every six inches using a surveyor's transit and a precision machinist's scale. With care, you will find it possible to place a line of center punch marks along the desired parabolic curve with a tolerance on the order of 0.005 inches (±). Figure 13-10 shows this technique.

To begin with, the anchors are driven in the concrete and the wires representing the focal axis and latis rectum are stretched out using the surveyor's transit. The squareness of the layout may then be fine adjusted using the 3, 4, 5 rule. For the latter, a distance which is some multiple of 3 is laid out on one wire and a distance which is a multiple of 4 is laid out on the other. When the wires are perfectly perpendicular, the distance between these points will be a multiple of 5. This test can actually be used to improve the precision of the layout.

In the next step, when the wires have been strung as nearly perpendicular as humanly possible, the mold sweep frame is laid upon the floor and jockeyed into position and lag-bolted to the floor. The aluminum sheet from which the parabolic track is to be cut is next jockeyed about on the frame until the vertex and the sail tip location are well up on the sheet. The sheet is then firmly screwed

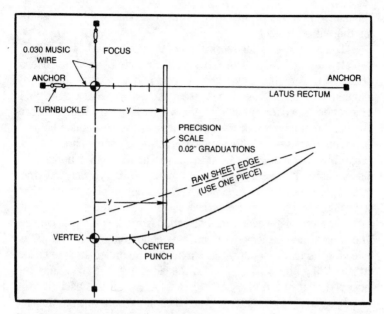

Fig. 13-10. Laying out the parabolic track.

to the mold sweep frame. Whereas the lag-bolting of the frame to the floor is a temporary attachment, the attachment of the aluminum sheet to the frame is permanent so using epoxy in the attachment is not a bad idea.

The vertex is next marked upon the sheet, and the string of points marking out the paraboloid are next carefully measured and punched with a centerpunch into the aluminum. On a large sheet such as this, it is advisable to use a blue toolmaker's ink (Prussian blue) to aid in finding and seeing the marks. When the entire string of marks has been punched in, the parabolic curve may be faired in between the marks. At this point it is not unlikely that one or more punch marks will be found that will not fair smoothly into the pattern. If this is the case, go back to the measurements and determine which points are spurious and correct them. This is especially important since any errors at this stage will be reflected in the entire final product.

Once a good true and fair curve is scribed upon the surface, the curve may be cut out using a sabre saw operating at slow speed. The saw kerf should run about 1/32 of an inch on the waste side of the curve. The curve may then be *carefully* hand filed to the line.

If you are not familiar with filing aluminum, it is a very good idea to obtain the assistance of someone who can tell you how. Get a set of new files for this task and keep the file card handy to prevent "pinning" of the files. The curve being generated will be the guide for the entire project.

The next step is the fabrication of the router cart. This unit is depicted in Fig. 13-11. The cart simply consists of a pair of rollers designed to run smoothly and with little play in a track formed of aluminum channel. The rollers are attached to a back plate and bracket designed for mounting a heavy duty industrial router. After this has been fabricated, the attachment of the tracks can begin.

The tracks—like the parabolic template itself—should be a single piece of aluminum (if at all possible) to avoid a noticeable discontinuity at the joints. With the parabolic curve finished, the tracks are arranged on the sheet and clamped in place in position such that the tip of a carbide ball cutter used on the router will extend a fixed distance in the X direction from the parabolic curve. The tracks may be jockeyed until this condition is obtained. The holes for the track mounting clips may then be drilled and the track attached. If a little "slop" is left in the clip holes, the track may be jockeyed a bit in the final touch-up adjustment of this assembly.

It is usually a good idea to depend upon the bottom track for the determination of the system position. There will usually have to be a little play in the tracks to prevent the cart from binding. If the bottom track is precisely located, the mold may be accurately cut despite this play by pressing down on the router.

It is important to note that the measurement to the ball cutter should be made in the X direction and *not* normal to the parabolic curve. As the router travels out from the center, it will tilt as shown in the illustration. When the track is properly installed, it will not be precisely parallel to the parabolic template because of this tilt.

With the track installed and the cart running smoothly, the next steps are in order. The first item is the attachment of a precision level to the sweep parallel to the latus rectum. The second step is the fabrication of the sweep bearings. These bearings may be made of hardwood and they should attach to the sweep so that they are offset the same distance from the parabolic curve as the center of the router. These bearings should be bored on a lathe for a smooth running fit on the pivot post. The final step is the attachment of the outboard roller on an adjustable bracket to the end of the sweep so that the roller can be made to rest upon the circular track. This roller should be cylindrical, and its axis should be as parallel to the latus rectum as possible, since an endward travel upon the circular level track is inevitable and an inclination in the axis of the roller will make for a scolloping of the reflector surface. With this installed, the mold sweep is complete. If the axis of the outboard roller and the pivot axis are both offset from the mold by an amount equal to the centerline offset of the router cutter, the tip of the cutter will describe a paraboloid of revolution as it is walked about the circular level track.

LEVELING

At this point the mold sweep may be unbolted from the floor and lifted into position above the mold. The pivot may then be inserted and the sweep locked onto the pivot with its bearings. The pivot collar is next set so that the sweep will clear the built-up blockpile of the mold.

Using a precision surveyor's level, the roller is next adjusted so that the difference in height between the vertex punch mark is close to the difference in X coordinate difference between the most outboard punch mark on the parabolic curve. The level is then unlocked and jockeyed to read level and locked in place. For a check, the sweep may be walked about the circular track 180° and the bubble level checked. If the bubble is off, it should be reset for half of the error and the roller adjusted for the other half. The sweep is then walked back to the first position and the height difference checked. This step is important since it determines the focal length of the reflector. Any significant departure will result in a non-parabolic surface.

FORMING THE MOLD

With the level securely locked and tracking around the entire circle, the task of cutting can begin. The collar can be lowered and

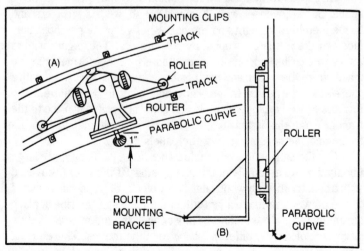

Fig. 13-11. Router cart at A and track at B.

the roller adjusted to give a level bubble reading with some of the block corners within reach of the router cutter. The sequence then consists of crawling about the surface of the mold pile and knocking off corners of the blocks. This part of the operation is relatively tedious and time-consuming but necessary. It is probably a good idea to have a stock of carbide ball cutters since several will be worn out before the mold is finished. Depending upon the care with which the stack was built, there may be a great deal of cutting to do in certain areas. It is probably a good idea not to try to take off more than about 1/4 inch in these initial cuts. In the initial cutting, the best tack is for the operator to climb up on the mold stack and work the router radially. As the job progresses, the mold will become smoother and smoother and it will be difficult to sit on the inclined surface without falling off. At this stage of the finishing, it is generally easier to lock the router at some fixed radius and walk the tail end of the sweep in a circle. When the cutting has proceeded to produce a smooth paraboloid of revolution, any voids which have been exposed may be filled with auto body epoxy or other filler and a final pass taken on the surface. After finishing the surface should be sealed with shellac or other wood sealer.

ASSEMBLING THE FRAMEWORK

With the mold complete, the task of making up the frames can begin. The first task is that of assembling the central core. The frames for this unit can be built upon the floor and the crossframing and wire bracing installed later.

At this point the mold sweep should be removed and the pivot should be removed as well. Figure 13-2 shows the stringers and frame members straight and the curve between the parabolic surface and the framing occupied by cut formers. These cut formers attach to the ribs and frames. The outlines for these formers may be traced from the sweep and the formers sawed out of plywood with a sabre saw. The central frame is now hoisted up into position and the formers braced against the mold and glued and nailed in place to the frame. When this is finished, the clevises can be installed on the frame and the construction of the panels begun.

For the panels, it is advisable to provide a flat floor space large enough for the assembly of the intercostals. If the floor is covered with a construction plywood false floor, blocks of wood can be nailed to the plywood to make a jig to hold the pieces of the ribs in place during gluing and assembly. In most cases the intercostals will be larger and of slightly different shape than the frames. Therefore, a second set of blocks can be used for jigging these. It is usually

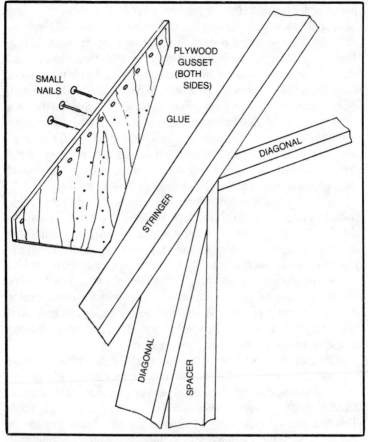

Fig. 13-12. Stringer spacer, diagonal, and gusset assembly.

worthwhile to make all of the ribs and all of the intercostals at one time since the duplication will provide certain economies of scale. All of the stringers, spacers and diagonals can be cut at a single saw setting and so on.

If the mass production of the frames is to be attempted at a single setting, it is usually worthwhile to build a dummy plywood template of each first so that the design can be checked for any gross errors before all of the pieces are cut and the error is replicated throughout the structure. A single dummy for each separate rib and intercostal design will suffice.

With the pieces cut, the stringers may be laid into the jig and the braces and diagonals glued in place and the gussets glued and nailed on one side. The frame may then be turned over and the gussets glued and nailed in place on the opposite side when the glue

223

has dried for the first. During the drying process, the members may be held in place with the sandbags prepared for the mold layup.

The basic construction of the rib joints can be seen in Fig. 13-12. The stringers, spacers and diagonals are fitted together and glued. A plywood gusset is glued and nailed to each side of the joint. Care should be taken to use small nails in this process. The strength comes from the glue and *not* from the nails. The glue should be a good waterproof glue made for marine or aircraft use. Strongly resist the temptation to drive nails into the endgrain of the spacers and stringers. In general the nails should disturb the grain of the stringers as little as possible, and the end nailing into the spacers and diagonals will only *weaken* the stringers!

Once a set of ribs and intercostals are assembled, it is possible to build a panel. To do this the formers are cut out and clamped to the ribs. The ribs are then temporarily clamped in place on the mold using the central core clevises and a falsework jig made to hold the ribs and intercostals in place. The formers are then glued to the ribs and intercostals using light nails as with the gussets. The formers should be pushed down into intimate contact with the mold surface during the gluing process. A sheet of waxed paper can be placed between the formers and the mold to prevent inadvertent adhesion to the mold due to spilled or running glue.

With the formers in place it is possible to begin adding the cross members. Figure 13-13 shows the means of attaching the crossmembers. An angle bracket is bent up of aluminum alloy and is attached to the cross member and the spacer with through bolts. This bracket not only serves to stiffen the joint but it provides a mounting place for the brace wire in the plane of the cross member (these are the wires running, for example, from E 13 to D 14, etc.). A second plate with turned-down ears is arranged to bolt under the corner bracket matching holes R and S. This plate is attached through the gusset with through bolts at T and U. This plate serves to anchor the diagonal wires; those running, for example, from E 13 to A 16, etc. Note that the bolt holes *do not* pass through the stringer but rather pass below it. Again, this is done to minimize the weakening of the stringer due to holes and fasteners. With these brackets in place, the cross members can be bolted in place and the brace wires taken up comfortably snug.

In order to insure a uniform dress of the brace wires it is nearly imperative that a wire tension gage be employed, such as the one shown in Fig. 12-14. It may be purchased from an aircraft supply house or fabricated rather easily. The gage operates on the principle that the pressure of the spring deflects the wire by an amount

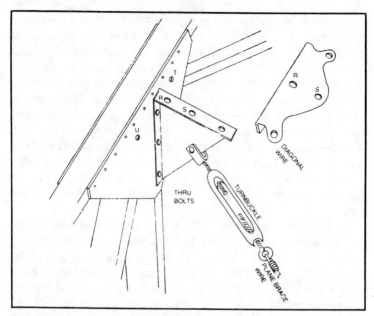

Fig. 13-13. Crossbrace and wire assembly.

proportional to the reciprocal of tension in the wire. The pulleys should be made fairly large in radius so that the stiffness of the wire does not contribute errors. The homemade gage can be calibrated by hanging known weights on a vertically hung wire and noting the gage deflection. For the wires in an assembly of this proportion, a preload strain of about 25 lbs is adequate to insure that the wire is not slack. Too much preload simply places unnecessary stress upon the structure.

INSTALLING THE SKIN

Once the panel frames are assembled, the completed panel frame may be removed from the mold. The frame is now ready for the skin installation. There are a variety of materials which can be used for the skin, but one of the easiest to work with is expanded aluminum mesh. This is an aluminum sheet material which has been slit and stretched so that a series of diamond-shaped holes are presented between somewhat twisted square section wires. One of the chief advantages of this material is that it is stretchy.

The expanded aluminum mesh is laid on the mold over a layer of waxed paper to prevent adhesion to the mold. Because of the compound curve of the mold, it will not lie flat. This problem is cured by stretching the aluminum with sandbags hung with hooks to the periphery of the sheet until the sheet is stretched like a sweater to the contour of the mold. With the mesh stretched in place, the edges of the formers are coated with a thixotropic epoxy glue and the panel section carefully lowered back into position and left for the epoxy to dry. The tension on the mesh should be relaxed only after the curing of the epoxy is complete. The panel may then be lifted and the fastening of the skin secured by means of light aluminum wires passed through small holes in the formers and the holes in the mesh and twisted tight. This process should start at the panel periphery and proceed toward the center. When the mesh has been completely secured in this manner, the mesh can be trimmed to the outline of the panel. Care must be taken to avoid denting of the mesh.

In the event that the mesh does become dented or sags in some areas, the reflector can be placed back on the mold and the mesh can be tightened in given areas by twisting the connecting part of the mesh slightly with a pair of long nosed pliers. This is one of the basic advantages in using the mesh. The second advantage is the fact that the openings of the mesh will give a good grip for the mirror tiles.

As a given panel is finished, it is carefully marked to match the central core. The core is then rotated to the next panel position, the panel is removed, and construction proceeds on the next panel. Before removal of the panel from the mold it is a good idea to give the panel the required wood sealer and finish. The mesh is protected in this position.

While it is certainly possible to simply finish the wood of the reflector for outside usage and to rust protect the wire stays, a good practice would be to cover the outside of the backframe with aircraft fabric of some sort. This serves to improve the structure in several ways. First of all, it protects the wood and stay wires from the weather and the sunshine. Unlike a sailboat, the back trusswork is a rather complex structure to refinish annually and perhaps more

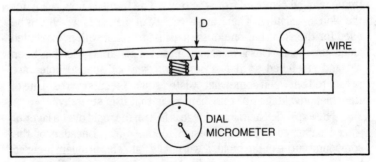

Fig. 13-14. The wire tension gage.

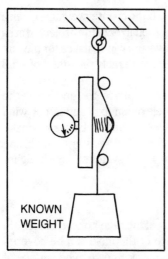

Fig. 13-15. Calibrating the tension gage.

KNOWN WEIGHT

often in very sunny climes. Secondly, the fabric enclosure will help by considerably reducing the wind resistance of the structure. This is advantageous both from the standpoint of windstorm survival and from the operational standpoint since a reduced wind resistance will make the problem of tracking the sun in high winds much less serious. There are a variety of fabrics with rated lives of 5 to 20 years in aircraft service. In general, the fabric is attached by both glue and sewing. The application of airplane dope will then shrink the fabric to a smooth drum-head consistency. A standard manual on aircraft covering should be consulted before starting this job.

When all of the panels have been finished, the central core can be surfaced with mesh in a similar fashion and the rear surface covered. It is probably advisable also to cover the interfaces of both the panels and the central core. In this way the structure of the backframe will be completely enclosed. This maximizes one of the advantages of the rear covering if the collector is to be used in a snowy clime in that the collector will accumulate much less snow, sleet, or ice on the smooth fabric backside than on the trusswork of the frame. It is important that the units be pretty well sealed since an amazing amount of snow can blow through a relatively tiny crack or opening in a storm with dry snow and high winds.

With all of the units finished, the next task is to weigh the panels and the backframe and to calculate the center of gravity of the structure. The reflector pivot and the counterweight can be made in the form of an "A" frame structure which attaches to four of the vertices of the central core, for example points 2 and 3 and 5 and 6 of Fig. 13-1. This structure can carry the azimuth axis and the counterweight that balances the reflector. If the counterweight is given a substantial flat area, equal to about one-quarter of the projected area of the collector proper, it will also serve as a wind balance for the unit in the azimuth plane and will reduce considerably the effect of wind torques upon the azimuth drive.

The collector servo drives can be made to function if the collector is not balanced about the elevation axis and is not wind balanced about the azimuth axis. However, these balances will make the achievement of a smooth acting drive much easier to attain and will permit operation of the solar collector system in much higher winds.

TILING THE REFLECTOR

With the mold available, the tiling of the reflector can proceed in one of several manners. The surface of the mold can be covered with the tiles laid with the reflecting surface down. The surface of the mesh can then be coated with a thixotropic epoxy cement, and the individual panels and the central core are lowered onto the glass and allowed to harden. This is the faster way to do the job than the individual tiling operation described in Chapter 11 for a ready-built reflector. However, it is often not quite so accurate as a well done tiling job where the tiles are individually set. In addition to this there is likely to be a fair amount of tile cleaning to do after the epoxy has set, unless the mix used is very non-runny. As before, the surface of the mold must be protected from adhesion.

JOINT EFFORT

As may be seen from the foregoing discussion, the construction of the mold and the reflector construction jigs and tooling is a lengthy and time-consuming process as well as being expensive. The tooling is capable of producing a number of reflectors. Therefore, it would make sense to do the tooling as a joint shared effort and build a number of collectors using the same tooling and building structure. As described and with some care in workmanship, the solar collector should have a useful life of 15 to 20 years with a reasonable amount of maintenance. With the expanded mesh backing, the reflector would also make an excellent satellite terminal or a radio telescope at times when the sun is not shining. Therefore, hobbyists with these interests could well be attracted into a joint effort to construct the reflector tooling. During the sunshine periods in the summer, the unit will cast a very large and fine shadow suitable for picnicking. This factor might be used to sooth your family's feathers during the initial discussions of the project.

14

The Solar Heat Engine

Having examined the various features and techniques for collecting solar energy to make some working fluid very hot we shall next take a look at the requirements of the system as a whole. A discussion of the properties of d-valve, corliss, and uniflo steam engines is beyond the scope of this text and is much better treated in a separate volume. As a practical matter there is no one in the United States regularly manufacturing such machines in the 3 to 10 horsepower range at this writing. There are a few manufacturers producing steam and gas turbines in this range, but these units are extremely expensive with prices quoted at $10,000 and above, which essentially removes them from consideration. In addition, a typical 3-hp steam turbine is about 6 inches in diameter and rotates at speeds on the order of 100,000 rpm. It would require a very expensive and precise reduction gear drive to get this speed down to one suitable for driving a 60-Hz alternator. From a practical point of view, the steam engine to be used in a solar home electrical plant will have to be converted from an available automobile or outboard motor.

By the time of the final flowering of the reciprocating steam engine, the engines of choice were *uniflow* types. These engines admitted the steam at the top of the cylinder and evacuated the steam at the bottom of the stroke through a series of ports surrounding the cylinder. This action prevented the alternate heating and cooling of the top of the cylinder and provided the longest possible conduction path between the relatively hot steam inlet and the relatively cool exhaust outlet port. These actions minimized the heat losses within the engine.

It is noteworthy that the common two-stroke outboard motor is provided with a similar set of ports. The bypass from the crankcase to the cylinder which admits the fuel and the exhaust port are both situated at the bottom of the cylinder as is the exhaust port of the uniflow engine. One can reason that the steam inlet port (which can be much smaller than the outlet port because of the higher driving pressure) can be installed in place of the sparkplug or the compression relief valve at the top of the cylinder. The principle missing item is the availability of a variable timing linkage to operate the steam inlet valve.

There are two difficulties with the conversion of an outboard motor to steam operation:

- The outboard motor is designed to be lubricated with a fuel-oil mist in the crankcase. For steam operation this mist will have to be replaced with a more positive lubrication.
- Outboard motors are designed without the oil-control rings which are used in a "wet" crankcase engine. In the outboard, it is only necessary to seal the compression since any oil on the cylinder walls will eventually be burned in any event. For steam operation some measures may have to be taken to reduce oil consumption.

Aside from these difficulties, the outboard would seem the most likely choice for a steam conversion unit. The 3600 rpm operating speed desired for a 60 Hz alternator is well within the normal operating range of these engines, and the pressures developed by the steam are much more modest than the sharp shock of the gasoline explosion.

Figure 14-1 depicts schematically the outboard-steam conversion. The piston is shown near the bottom of the stroke with the bypass and exhaust ports (which have both been connected to the condenser) open. It is of course necessary that the portion of the bypass leading to the crankcase be blocked. The upper end of the piston stroke is shown dotted. A steam valve has been inserted into what used to be the spark plug hole. You will note that the cone of the steam valve is reversed with respect to the cone used on automotive valves or the compression release valve usually used on outboards since the pressure is always higher on the steam supply side of the line than on the cylinder side. An oil-control ring has been added to the piston.

The normal reed valves have been closed off except for a small section which allows the oil mist lubrication of the engine through a

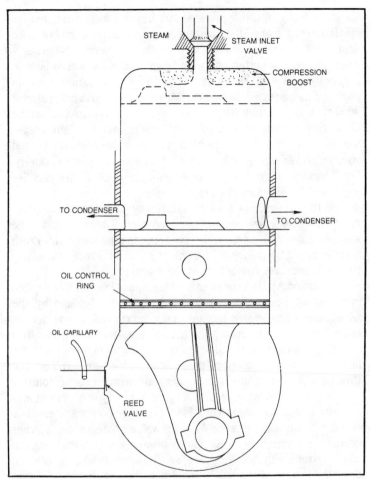

Fig. 14-1. The outboard-steam conversion.

capillary tube. Since the condenser will be operating well below atmospheric pressure, there will be a net leakage of air from the crankcase to the condenser and a small pulsating flow of air and oil mist at the top of each stroke will lubricate the engine and make up the loss. Both the air leakage and the oil pickup are probably unavoidable. These impurities will have to be removed from the condenser later. This is not an unusual operation since most steam engines had air and oil scavenging provisions for the condenser.

SPEED CONTROL

As had been noted earlier in Chapter 4, there are distinct advantages to operating the home electrical system at precisely 60

Hz since many of the appliances and devices were designed for 60-Hz operation and such matters as power factor correction are dependent upon the maintenance of the frequency. Whereas a windmill may be better operated to generate DC which is subsequently converted to 60-Hz AC, the hydro plant and the steam plant are more amenable to direct drive of the 60-Hz alternator at 3600 rpm. Because this type of operation is more efficient than the direct current generation/conversion process, it seems advantageous to consider the system built in this manner. Auxiliary and standby power can be obtained from batteries charged by conventional battery chargers, or from an automotive alternator rectifier mechanically driven by the steam engine.

In the presence of a widely varying system load, the requirement arises for a fast responding speed control mechanism for the steam engine. With a heavier load the engine will tend to slow down and the output frequency will fall unless a control system is available to rapidly increase the output of the machine.

Traditionally, the speed of steam engines has been controlled in power applications by means of a throttle valve and by the adjustment of the valve linkage. The throttle valve was usually controlled by the governor, and the valve linkage was generally a manually operated affair adjusted by the engineer to account for larger and slower variations in the load. On steam turbines, the throttle valve is essentially the only mechanism for speed control.

For the small home electrical system, this type of operation is somewhat less desirable than it is for the large power system. For one thing, the home electrical system will experience much larger changes in demand as a percentage of installed kVA than will the large system. Furthermore, the small system will generally not have the benefit of the pull-in capability of a large number of synchronized machines. The small home system must therefore be capable of sustaining a much faster response rate to changing load.

ELECTRONIC STEAM INJECTION

To meet these requirements with maximum efficiency, the writer is prone to favor the use of a very modern technique which is analogous to the electronic controlled fuel injection which has begun to make its appearance on sophisticated internal combustion engines. In Fig. 10-7, a rather idealized pressure versus volume curve for a vapor cycle engine was presented, and it was noted that the area under the curve represented the work done by the engine. It should be relatively obvious that lengthening the segment AB in this curve will increase the area of the curve. It also represents the consumption of more steam. The length of segment AB represents a measure of the length of time that the inlet valve was admitting steam to the cylinder. More steam equates to more work output. Similarly, shortening the curve segment AB would represent admission of less steam to the engine and therefore less work. An electronically controlled solenoid valve for steam admission could be arranged to permit control of the steam admission on a stroke-by-stroke basis and would thus provide the fastest response possible to changing load requirements for any given engine.

INLET VALVE TIMING

The older steam engines made use of a rather large ratio of connecting rod length to stroke or crank throw. If the connecting rod were infinitely long, the motion of the piston would be simple harmonic motion; that is, the graph of piston position versus time would be a sine wave. This has several advantages:

- The vibration moments would be confined to a single frequency equal to the crank rate.
- The piston would spend an equal time at each end of the cylinder. This was important in a double-acting engine in which both ends of the cylinder were used.

Modern internal combustion engines have gone in the direction of the use of a very short connecting rod compared to crank throw since this tends to make the engine much more compact. Furthermore, no modern high-performance engine makes use of a double-acting design. The harmonics introduced by the non-simple-harmonic motion of the piston have been canceled by the introduction of sophisticated balancing techniques.

A difference is brought about in the "indicator diagram" or P versus V diagram of the engine by this design change. Figure 14-2 shows an indicator card diagram more typical of the results to be obtained for an outboard-steam conversion. The horizontal axis has been plotted to show no scales. One of these is the percentage of full cylinder volume and the lower one is the angle of crank rotation. From the second curve it may be ascertained that the engine spends a good deal of its time near minimum volume and even more of its time near maximum volume with the period in between being traversed very rapidly. The rotation curve was calculated assuming that the engine has a compression ratio of 6:1 and that the connecting rod length is 2.5 times the crank throw radius. These figures are representative of modern high-performance internal combustion engine design.

The figure shows one curve in full representing the nominal load indicator card. Also shown dotted are two curves representing

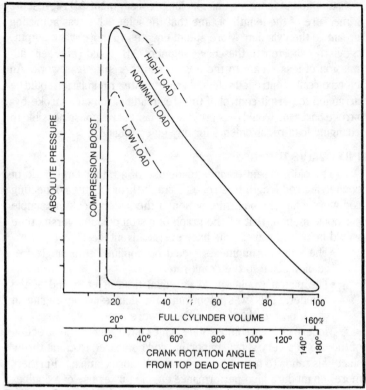

Fig. 14-2. Outboard-steam conversion indicator cards.

low and high load operation. In the low load condition, the inlet valve was open for such a short period that the pressure in the cylinder did not have time to approach the boiler pressure and the valve is performing a time modulated throttling action. At the nominal load condition, the valve was open long enough for the cylinder pressure to nearly reach boiler pressure. In the high load condition, the valve was held open long enough for the engine to hold near boiler pressure in the cylinder for about 35° of crank rotation after top dead center.

If we assume that the engine is turning at 3600 rpm, then the crank is making a full revolution every 0.0167 seconds (or 16.7 milliseconds) in the low load condition, if we wanted to have the inlet valve open 10° before top dead center and reclose about 15° after top dead center (TDC). The valve would be open for 1 millisecond. At nominal load, if we wish to open the valve 15° before TDC and reclose it 25° after TDC, the total period would be 1.85 milliseconds.

234

This is a very fast action indeed, being comparable to the opening time of a high-speed camera shutter. However, similar performance has been attained in cold-gas thruster valves employed in the maneuvering mechanism for spacecraft.

A valve to operate at such speed can be constructed with an ordinary poppet type mechanism driven by a moving coil solenoid of the type used in loudspeakers, but this would require a considerable amount of electrical energy. Figure 14-3 shows a servo type valve which can be operated with considerable less electrical drive.

In the servo valve, a steam inlet enters a cylinder containing a spool. The body of the spool is shown blocking the outlet, held in place by a spring. Since the steam pushes equally on both ends of the spool, the spool is in equilibrium. If the small solenoid marked *solenoid 1* is briefly pulsed, the needle valve releases the pressure in the bottom chamber and the valve spool will snap down to a new equilibrium position with the steam outlet open. *Solenoid 1* may then be reclosed. When the time arrives to re-close the inlet valve, *solenoid 2* is briefly opened and the pressure in the upper chamber falls. The restored pressure in the lower chamber overbalances the reduced pressure in the upper chamber and flings the spool up again. In effect the valve itself is a small free-piston steam engine. If the pressure releases are placed somewhere up from the end of the

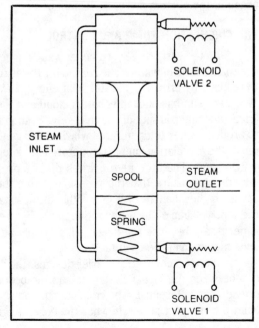

Fig. 14-3. The servo inlet valve.

piston, as the spool passes the release outlet, the remaining steam is trapped in the end chambers. The compression of this steam prevents the spool from slamming violently into the end walls of the chamber. Naturally, the spool must be precisely lapped to fit the valve cylinder in order for the valve to operate without much leakage.

Another item is worthy of attention. An ordinary auto or outboard engine has the compression ratio limited in order to prevent detonation or knocking in operation. No such limitation exists in the steam engine. The *clearance volume* at the top of the cylinder simply absorbs steam. Since the amount of steam in the cylinder is larger, this would tend to make the diagram fatter. A closing off of some of this clearance volume by the addition of a block to the cylinder head will improve the operation of the engine with respect to the maximum to minimum output ratio and will improve the efficiency of the engine, particularly under light loads. The reduction of clearance volume is best accomplished on the cylinder head where it does not affect the operation of moving parts and upset the balance and vibration damping of the engine.

THE SOLAR POWER SYSTEM

The complete solar power system is depicted schematically in Fig. 14-4. We shall briefly discuss each of the items in terms of usage in the system and requirements.

THE CIRCULATION PUMP AND CONTROL

As noted earlier, the solar boiler was designed to contain a very minimal volume of water. The reason for this is to permit the boiler to respond by providing useable heat energy in very brief periods. When the sun has ducked behind a cloud and the boiler cools, the circulation pump is shut off. Conversely, as soon as the boiler starts to produce water temperatures which would contribute to the flash tank, the circulation pump is operated to keep the boiler from overheating. The little item marked dP is a pressure constriction intended to ensure that the pressure within the boiler is always slightly higher than the pressure within the flash tank. This prevents the actual boiling of the water within the boiler proper. This addition is necessary because the boiler will probably be installed at some level higher than the flash tank.

In more common steam boiler designs, the difference in elevation between the higher flash tank and the boiler tubes is used to provide this differential. The circulation pump will operate continuously but at a very slow rate when the boiler is producing no heat.

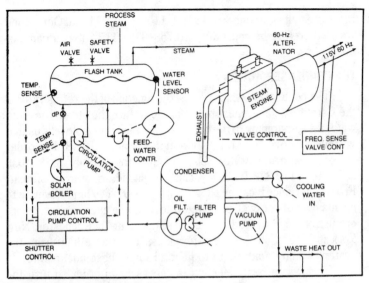

Fig. 14-4. The solar power system.

THE FLASH TANK

The flash tank serves the function of storing the hot water generated by the solar boiler and separating out the steam from the superheated water. The water enters the flash tank in the liquid state under high pressure. When the engine draws some steam from the flash tank, the pressure falls slightly, permitting some of the superheated water to flash off into steam. If this tank is made small, the system will come up into operation very rapidly. However, the system will have little stored energy to coast down, if the sun passes behind a cloud. On the other hand, if the tank is made large, it can store a considerable amount of energy. This comes at the expense of a lengthy time to get up-to-steam. If the up-to-steam time is too long, the ability to use brief intervals of sunshine is lost since the system may not get up-to-steam during the brief interval and no power may be produced at all.

For a numerical example, let us consider a system which delivers 2.24 kW of electrical energy. Let us further assume that the losses due to mechanical inefficiency of the engine, the water pumping, and the tracking of the solar reflector bring the efficiency of the unit down to 0.71. This brings the useful power requirement up to 3.15 kW or 10,768 BTU/hr. If the inlet to the engine has saturated steam at 200 PSIA and the outlet to the condenser is at 1 PSIA, we may refer to the saturated steam information in Table 14-1 and observe that the total heat difference between saturated steam at

237

these pressures amounts to 1198 − 1105 = 93 BTU/lb. Our engine would therefore be consuming 10,768/93 = 115.8 lb of steam per hour.

THE CONDENSER ACTION

The condenser, which serves the function of the refrigerator in the Carnot machine, must be prepared to turn the steam back into water for re-pumping to the boiler. In this case, the table tells us that we must remove 1035 BTU/lb from the steam at 101.8° F in order to condense it into water. This amounts to 115.8 lb/ hr × 1035 BTU/lb = 119,832 BTU/hr. This is a significant amount of heat. From experience it can be shown that 80,000 BTU/hr is adequate to maintain a not-too-well insulated trilevel home at an internal temperature of 65° F when the external temperature in Rochester, New York is −15° F with moderate winds. The waste heat from the system would be adequate to heat a rather substantial home.

This waste heat rejection is necessary in order to turn the exhaust steam back into water for restoration to the boiler. If the engine were operated in a noncondensing fashion and allowed to simply exhaust to the atmosphere, there would be several sequelae, none of which are advantageous.

First, if the exhaust pressure were raised to 14.7 PSIA, we see from the steam tables that the available energy would be 1198 BTU/lb − 1150 BTU/lb = 48 BTU/lb. This is just a bit more than half of the energy attainable with the condenser. We would have to generate nearly twice as much steam.

Secondly, if the engine exhausted to the atmosphere, it would be necessary to provide valves at the top of the cylinder to vent the cylinder to the atmosphere. Otherwise the engine would give up even more of our hard-won solar energy in compressing the residual steam to five, six, or more times atmospheric pressure.

Thirdly, the exhaust noise would be a distinct problem and would require extensive muffling. Those old enough to recall the steam locomotives will adequately attest to the noise of a noncondensing steam engine.

Finally, the boiler water would be lost. Ordinary tap water and lake and river water contains far too much dirt and dissolved minerals to be run through the boiler untreated. On a small system such as this, it is practical to operate the engine on an initial charge of distilled water and to provide a distillation apparatus which can be easily cleaned of boiler scale to provide a continuing supply of makeup water.

The condenser can be designed to be either liquid or air cooled. For air-cooled designs, it is common to attain heat transfers of 11 to

14 BTU/ft^2/hr° F. Water cooled condensers are much more common and much more compact. In water-cooled designs it is common to achieve a cooling rate of 150 to 250 BTU/ft^2/hr/ °F. With the steam emerging 10 to 20°F above the water outlet temperature and a water temperature rise of 15 to 20°F. To achieve a steam pressure of 1 PSIA, we would require a water inlet temperature on the order of 70°F which is within the range of outlet temperatures for a radiant heating system for a home. Presuming that the temperature drop across the condenser tubes is 15°F and that we could attain a transfer rate of 200 BTU/ft^2/hr/°F, we would require a condenser tube area of 119,832 BTU/hr divided by 3000BTU/ft^2 = 40 ft^2. For ordinary half-inch copper water line, this would require about 270 lineal feet. The water would emerge at a temperature of 85 to 90°F which would again be within the range capable of providing good radiant heating for a home.

BACK TO THE STORAGE TANK

To provide this action, the boiler would have to provide not only the 10,768 BTU actually utilized in the engine but also the 119,832 BTU rejected by the condenser for a total of 130,600 BTU/hr. As a cross check we see that the 1198 BTU/lb contained in the 200 PSIA steam minus the 70 BTU/lb contained in the condenser water gives an addition of 1128 BTU/lb for the boiler. Multiplying this by 115.78 lb/hr yields 130,600 BTU/hr.

If we presume a boiler heat input (with the engine inoperative) of 130,600 BTU/hr and we presume that we would like to have the boiler up-to-steam from a starting temperature of 32°F in perhaps 15 minutes, we may calculate the amount of water in the flash tank. At 381.8°F we see that water has a heat content of 355 BTU/lb. Dividing this into a quarter of our hourly input gives an answer of 92 lb for the flash tank contents.

THE CIRCULATION PUMP

Next let us see what is required of the circulation pump. If we presume that we employ a pressure differential of 50 PSI between the boiler and the flash tank to prevent the boiling in the boiler proper, we see from the tables that we could add 21 BTU/lb to the water in the boiler. For the required heat transfer this would mean that we would have to pump 130,600 BTU/hr divided by 21 BTU/lb = 6,219 lb/hr through the boiler. Since water is incompressible, the work required is simply the force times the distance and the power required may be calculated:

$$\frac{50 \text{ PSI} \times 144 \text{ in}^2/\text{ft}^2 \times 6219 \text{ lbs}}{62.4 \text{ lb/ft}^3}$$

$$\begin{aligned} &= 717{,}582 \quad \text{ft-lb/sec} \\ &= 199.33 \quad \text{ft-lb/hr} \\ &= 0.36 \quad \text{hp} \\ &= 270 \quad \text{watts} \end{aligned}$$

The water would undergo a 19.2°F temperature rise.

THE FEEDWATER PUMP

The feedwater pump has to move the water from the condenser to the boiler against a pressure differential of 200 PSI. For this the calculation becomes:

$$\frac{200 \text{ PSI} \times 144 \text{ in}^2/\text{ft}^2 \times 115.78 \text{ lb/hr}}{62.4 \text{ lb/ft}^3}$$

$$\begin{aligned} &= 53{,}437 \text{ ft-lb/hr} \\ &= 14.84 \text{ ft-lb/sec} \\ &= 0.027 \text{ hp} = 20.4 \text{ watts} \end{aligned}$$

The feedwater pump which fulfills the compression portion of the vapor cycle requires a negligible portion of the cycle energy in a vapor cycle engine.

On the other hand it may be seen that the circulation pumping does require a significant amount of the energy. This is one of the disadvantages of a system design which requires that the heat addition take place at a point physically removed from the flash tank. This is probably a penalty which is difficult to avoid in a tracking solar collector system.

THE dP REGULATOR

From the foregoing discussion of the bimodal operation of the circulation pump, it should be relatively obvious that the item designated dP in the system diagram should not be just a simple constriction in the piping but rather should be a pressure regulator. Figure 14-6 shows one configuration of such a device. The pressure differential between the solar boiler and the flash tank presses against the diaphragm and compresses the spring, thus separating the moving jet from the fixed jet. As the water flow increases, the jets open wider and tend to allow more water to flow, thus holding the pressure differential across the regulator to a relatively constant level. For the idling condition the small low-pressure bypass permits a small amount of flow. The inclusion of this regulator provides for a smoother operation of the system.

THE AIR VALVE

The air valve shown at the top of the flash tank serves to admit air to the flash tank whenever the temperature in the flash tank falls

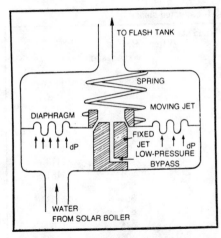

Fig. 14-5. The dP regulator.

below 212°F. This prevents the flash tank from being exposed to atmospheric crushing and water suction from pushing water slugs into the steam piping. At startup, this valve is held open until the water in the flash tank reaches atmospheric boiling temperature. Once boiling begins in the flash tank this valve closes. The steam will expel the majority of the air from the system.

THE VACUUM PUMP

The purpose of the vacuum pump is to remove the air and gasses dissolved in the boiler fill. This unit operates only periodically. It must be capable of developing a vacuum equivalent to the best condenser vacuum in order to function effectively.

THE SAFETY VALVE

The purpose of the safety valve is naturally to limit the pressure within the flash tank to the design level for the system. This valve should be designed following the ASME Boiler Code or some overriding legal code. The code specifies not only the opening pressure but also the size of the valve outlet. The safety valve is a most important part of the system. Without the safety valve, the flash tank is a timebomb!

GENERAL

A few general notes are in order on the subject of the system. First and foremost of these is the fact that the various pumps, tracking motors, and control apparatus should be arranged so that they operate on the auxiliary DC storage power. This is necessary to permit system startup. The system cannot be operated without

Table 14-1. Saturated steam.

Absolute Pressure PSIA	Temperature °F	BTU/lb above 32°F In Water	Heat of Vaporization	Total Heat
1	101.8	70	1035	1105
2	126.1	94	1022	1116
5	162.3	130	1000	1130
10	193.2	161	982	1143
14.7	212.0	180	970	1150
20	228.0	196	960	1156
50	281.0	250	923	1173
75	307.6	277	904	1181
100	327.8	298	888	1186
150	258.4	330	863	1193
200	381.8	355	843	1198
250	401.0	376	824	1200
300	417.3	394	808	1202
400	444.6	424	780	1204
500	467.0	450	754	1204

these items and therefore cannot depend upon running system output.

Secondly, the system piping should be well insulated on all of the hot lines in order to maintain some reasonable standards of safety and efficiency.

Thirdly, the safety aspects of the design of such a system cannot be overemphasized. At temperatures approaching 400°F, steam can melt human flesh from the bones in seconds. The design of even such a small scale system requires the exercise of very good mechanical engineering practices and should not be attempted by amateurs.

OTHER HEAT ENGINES

With reference back to Chapter 10, we see that any type of heat engine which has the compression, heat addition expansion, and heat rejection cycle could be used. One unit operating on the sterling cycle principle with the heat receptor located at the focus of a small paraboloid is commercially available. However, the size of this unit places it more nearly in the class of a toy than of a practical unit.

In principle, a gas turbine system which compressed air in a rotary compressor, heated it in the solar boiler equivalent, and then expanded it back to the atmosphere or to a condenser equivalent could be made to operate with reasonable efficiency. However, the cost of development of such a unit is prohibitive for the individual. As a practical matter, the use of solar steam appears to be the most practical choice at this writing.

15

Alternate Energy
Storage Techniques

As noted in Chapter 3, the only really economically viable means of storing any significant quantity of electrical energy at the time of this writing is by means of lead-acid storage cells. Unfortunately, these leave a great deal to be desired in terms of capital investment, maintenance, and life-cycle replacement cost. Despite these factors, the lead-acid battery is still used by telephone companies and many computer installations for standby power because of lack of a viable alternative.

Few of the power companies have any significant means of storing excess energy. Those few that do, accomplish the storage by running a hydro plant in reverse to pump water into the impoundment. Typical overall efficiency for such operations is approximately 60 percent. The pumped water storage is really a very satisfactory system provided there is room for the impoundment. Probably one of the best and most trouble-free home electrical installations would consist of a series of water-pumping windmills feeding an impoundment and a hydroelectric plant drawing water from the impoundment. This alternative is of course available only to those homes with suitable amounts of real estate to contain the impoundment and with ready sources of water. The very considerable advantages of a hydro installation were noted earlier.

HYDROGEN

One of the commonly discussed mechanisms for storing of excess electrical energy from a fluctuating source entails the release of hydrogen from water by electrolysis. This is a very efficient process in terms of BTUs of fuel content versus overall electrical input. In addition to this, hydrogen is a nearly ideal fuel. Hydrogen will burn without any soot or residue and if burned with the oxygen also liberated in the electrical process, the only combustion product is water vapor. If the hydrogen is burned in air with a slight excess of air under high temperature and pressure as in an internal combustion engine, there can be some NO_x formation; however, if a slight excess of hydrogen is present this would not be the case. Hydrogen is also about the most energetic fuel there is, with a heat of combustion of about 52,000 BTU/lb. This is more than 2.5 times the energy of gasoline, 3.5 times the energy of coal, and 7 times the energy of wood.

Unfortunately, aside from the "Hindenberg Syndrome," there are some very real problems associated with the handling and use of hydrogen as a fuel. Hydrogen weighs only 0.0056 lb/ft^3 at standard temperature and pressure. It would therefore require a volume of 179 ft^3 to contain a pound at standard temperature and pressure. An internal combustion engine which is very efficient will run on about 0.24 lb of hydrogen per horsepower-hour. Our three horsepower example would therefore be consuming something on the order of 129 ft^3/hr of hydrogen, and a 24-hr run would require about 3096 ft^3. The gas could be stored in a balloon; however, extreme caution would have to be exercised to keep all oxygen out of the ballon to avoid a "Hindenberg reaction."

Of course, hydrogen is used as a propellant in some of our highest performance rockets; however, it is used in liquified form. There are a few problems with liquefaction of hydrogen. At room temperature, liquid hydrogen has a vapor pressure on the order of 10,000 PSI. Vessels and plumbing to contain it would have to be immensely strong and tight. On the other hand, the boiling point of liquid hydrogen is −422.9°F. An open vessel of cryogenic liquid hydrogen will maintain itself at this temperature by boiling off slight amounts of the liquid. The space vehicles use hydrogen in this cryogenic form. Unfortunately, this is not all cakes and ale either. The hydrogen which boils off to keep the temperature down must be allowed to escape the vessel. If vented to the atmosphere, it will rapidly rise and dissipate; however, the vicinity of one of the tanks is a definite no-smoking area. Furthermore, most of the common materials become brittle as glass at these low temperatures. It seems likely that cryogenic liquid hydrogen will remain a fuel to be manipulated only by experts in high-technology applications.

Probably the greatest hope for the use of hydrogen as a fuel is the possibility of the development of a solvent-carrier technique in which the hydrogen is dissolved in a fluid the way that carbon dioxide is dissolved in soft drinks or beer. Such a technique is currently widely used for acetylene shipment. Experiments in which hydrogen has been dissolved in lithium hydride and other chemicals have been reported, but at this writing no such process or apparatus exists on a commercial scale.

WOOD AND VEGETATION

Probably the easiest means of energy storage available to the home electrifier is simply the growing of wood or vegetation which can then be burned. One of the great advantages of the solar steam installation is the fact that the system can easily be modified to employ a conventional fuel-burning boiler to provide steam for the system when the solar energy is not available. Wood will provide between 7,000 and 8,000 BTU/lb depending upon the type. Unfortunately, wood and foliage tend to burn in the initial damp phases with a great deal of smoke. One possible solution to this problem would be to make use of the excess steam produced during periods of high solar flux to drive off the moisture from the vegetation or wood.

If the drying is performed in a sealable vessel with the steam tubes on the outside, the process can proceed in three steps. In the first step, the wood or vegetation can be brought up to some temperature like 220°F with the vessel vented to the atmosphere. After a holding period at this temperature, the overwhelming majority of the moisture will be driven off. After this, the vessel may be sealed from the atmosphere and the temperature taken up to near the boiler temperature. During this period, the wood or vegetation will emit a volatile combustible gas. This gas is directly suitable for use as a fuel in the boiler or in an auxiliary internal combustion engine. Depending upon the temperatures achieved in the retort, the wood or vegetation residue in the retort may be entirely or partially converted into charcoal which makes a fine hot-burning smokeless fire for heating boilers (or for blacksmithing or broiling steak). A portion of the initial drying could of course be accomplished simply by placing the material in racks above the ground exposed to the sunshine. However, except in very dry climates, the drying will not be very complete, and the initial gas produced from the retort will have a large moisture content.

It should be noted that the gasses produced by this process are explosive and the venting should be accomplished through a suitable flame-holder to prevent an explosion in the retort. Unless a tall chimney is available for obtaining a natural draft, the charcoal fire will have to be supplied with a forced draft from a blower if a sufficient fire for steam production is to be maintained. The charcoal product obtained from this process has the advantage that it may be stored nearly indefinitely without loss and is always ready for energy production. Storage should be in a dry place or shelter.

DIRECT HEAT STORAGE

In the previous discussion, the flash tank was considered to be about minimum in size. It has been suggested that a very large flash tank could be employed to store heat directly. This tank could be buried in a large excavation and insulated with some high temperature insulation material in a very heavy layer. The principle difficulties with this type of scheme are the problems involved with keeping the insulation dry so that the insulation value would be retained and the cost involved in building a large high-pressure vessel. There is an advantage in scale in schemes of this sort since the surface area of a vessel tends to grow as the square of the radius, whereas the volume of the vessel tends to grow as the cube of the radius. This means that a large vessel will suffer a smaller percentage of heat loss than a small vessel. The required wall thickness also tends to grow as the square of radius, whereas the energy storage capacity tends to grow as the cube of radius. Naturally, a large storage system which contained enough energy to permit a coast-down time of several days to a week would come up to temperature slowly as well and might require a time on the order of weeks to a month to get up-to-steam from an initial cold start.

COMPRESSED AIR

Another energy storage scheme which has sometimes been suggested makes use of compressed air pumped into a cave or a mine. Aside from the problems of leakage, this scheme suffers from several problems. First of all, the compressors which are used to pump the air in the first place tend to deliver the air hot; that is, a significant portion of the work in compressing the air is contained in the heat. Unless the cave was well insulated, this heat would eventually leak off into the soil and would be lost. Upon emerging from the cave and being expanded, this air would become quite cold in the engine; thus, creating a further inefficiency. Of course, the cooling action could be used for refrigeration or air conditioning. The second problem is perhaps more serious. There are a number of locations where there is a small continuous leakage of natural gas from the depths of the earth. In a large cave, this leakage could eventually reach explosive proportions. In addition, it is very difficult to operate a large compressor without the entrainment of a certain amount of oil vapor in the output. For either or both of these causes, the development of an explosive mixture within the compressed air is a distinct possibility. The creation of a spontaneous spark due to rocks falling from the ceiling or some other cause is also a distinct possibility. Unless great care is taken in monitoring the air for explosive vapors or gasses, the possibility of a huge explosion exists.

SUMMARY

For those having the geographic advantages and the real estate to make it possible, the pumped water energy storage is probably the best available. For the rest of us, the consumption of wood or combustible vegetation seems to offer the cheapest and most viable form of energy storage. Future developments in vehicle-solute technology may eventually make hydrogen storage the best and least expensive of all.

16

Summary

The engineering study which led to the production of this book was intended to develop some basic points:

- A determination of the economic and engineering feasibility of the production of home electric power by the individual.
- Development of a list of practical resources for acquisition of the necessary apparatus.
- Development of a group of design rules for the construction of a practical and effective system.

In keeping with these goals, a series of nearly 70 letters were sent to manufacturers who advertised equipment which might be suitable or could perhaps be modified to be suitable for home electricity production. These letters spelled out in detail the requirement for equipment in the 2 to 25 kVA range and were tailored to the product of the manufacturer. The responses were sparse. A few of the manufacturers of high-speed turbines kindly responded but noted that their products were high precision items intended to operate fuel pumps on guided missiles, provide propulsion for torpedos, or drive spin testing machinery. Not a single response was obtained from any manufacturer indicating that a reciprocating steam engine was available in the appropriate size range.

A similar lack of response marked the investigation into the small hydraulic turbines. While there have been references to turbines manufactured in Europe, a catalog sheet, price information, and engineering specifications for suitable units were not submitted.

In the area of windmills for electrical power generation, there are several units manufactured in the US and in Europe which are capable of producing nearly a kW under high wind conditions; however, the average production for such units in even relatively windy areas (9 mph avg.) falls to such a level that they would be capable of powering a desk lamp and *perhaps* a radio but *not* a color TV.

The conclusion to be drawn from this investigation is that manufacturers have not yet found it economically feasible to address this market. At the time of this writing, it seems that the would-be home electrifier is going to have to build the overwhelming portion of the apparatus himself.

In the areas of DC to AC conversion, and computer load control, the picture is somewhat brighter since commercial equipment is available, albeit expensive. A number of vendors supply these items with suitable ratings. On the other hand, no manufacturer seems to supply automatic power factor correction apparatus.

Throughout the study, the outlook has been maintained that a home electrical system should be capable of doing the entire job and supplying all of the electrical needs of the family. A system which will take care of 10 percent of the requirements in a modern home was considered to be trivial because it is expensive to install and would not be economically justifiable: It would only slightly reduce the electric bills. This might be an interesting hobby item and would certainly provide a conversation piece, but it can in no sense be described as practical. As nearly as it is possible for me to determine, there is not a single home in the US which can be described as a modern electrified home which is running completely independent of the utility companies (excluding diesel motor generators).

Perhaps the most surprising part of the study has been the extent to which Americans have become energy consumers. The picturesque and huge German and Dutch wind-powered sawmills actually produced no more power than a fair sized chain saw which can be lifted with one hand. The big, old water wheel that ran the flour mill and ground grain for an entire village produced less power than the typical American family uses with only their air conditioner and vacuum cleaner. The maintenance of our standard of living requires a lot of power.

As a result of this, it may be seen in the preceeding chapters that only a few people could hope to provide home electrical power at a net economic savings. If one lived atop Mount Washington where the wind averages 35 mph, a rather modest windmill would suffice to electrify the home.

Similarly, someone living in a mountainous area in the vicinity of a substantial waterfall could have hydroelectric power at an economic saving. The dweller in Arizona or New Mexico or other sun-drenched areas could get by with a minimal sized solar collector on the order of 22 feet in diameter. However, the lack of availability of Fresnel lenses or mirrored reflectors of suitable size and the costs of steam plant installation and maintenance would make this a questionable financial investment.

This entire picture is of course predicated upon the concept that cost of electrical energy from the utility companies will grow no faster than perhaps a half percent higher than the general inflation rate. A faster growth rate in utility prices would of course make the investment in a home electrical generation system more attractive. Time alone will tell in some of these matters. Conceivably, a rising growth rate in utility prices may also bring about a change of zoning requirements so that we may begin to see 22-foot windmills or 32-foot solar reflectors rising above houses on suburban lots. Unfortunately, I fear that neighbors are not ready for them right now.

Index

Edited by Raymond A. Collins